PRAISE FOR LENORE NEWMAN

"Newman's jaunts through the animal kingdon[...]med meals with her friend Dan as she ponders how historical extinctions are linked to our current food systems, what we can do about it, and how humans must follow the example of the famed New York 'pizza rat,' and adapt to the food that comes their way." — *Booklist*

"Free-wheeling look at the flora and fauna we've eaten into oblivion."
— *Toronto Star*

"Edifying and entertaining . . . Never didactic and cautiously optimistic, Newman recognizes that there is hard work ahead to recalibrate the North American diet. She builds a compelling case for us human super predators to rethink our food choices, and to be healthier for the environment and our fellow inhabitant species. *Lost Feast* is enjoyable reading about a serious topic." — *Foreword Reviews*, starred review

"An interesting and thought-provoking adventure alongside an engaging, wry-humored narrator, the book forces the reader to consider humans' role in historic plant and animal extinctions, as well as how we might approach food more reasonably moving forward." — *Civil Eats*

"*Lost Feast* is buzzy, compelling, and genuinely funny."
— *Literary Review of Canada*

PRAISE FOR EVAN FRASER'S *EMPIRES OF FOOD*

"Spanning the whole of human civilization, this is a compelling read for foodies, environmentalists, and social and economic historians."
— *Kirkus Reviews*, starred review

"*Empires Of Food* will forever change the way you think about the contents of your shopping basket. At one level, [it] is a warning to all of us to shop and eat more responsibly. It is also a richly entertaining history of our relationship with the food that we put on our plates. —*The Express*

"Forget the old stages of human history, the familiar stone, bronze, iron age sequence: . . . Fraser and j . . . Rimas make a convincing case that food—or rather, food surpluses—best explain the rise and fall of civilizations. — *Macleans*

PRAISE FOR EVAN FRASER'S *BEEF*

"[A] cross-cultural survey of the omnipresence of cattle in myth, religion, art, history and culture, from the Lascaux cave paintings to pampered Japanese Wagyu cows." — *The New York Times Book Review*

"[L]ively and unsettling history-cum-polemic . . . they write vividly . . . and they certainly aren't antimeat; their colorful account is well-seasoned with a series of 'culinary interludes'" — *Publishers Weekly*

". . . *Beef* . . . reminds us that as tasty as burgers and steak may be, there's a price to be paid — in oil, land, and treasure." — *Time*

"An ambitious cultural-historical-agricultural history . . ." — *The Guardian*

DINNER
ON
MARS

**THE TECHNOLOGIES THAT WILL FEED
THE RED PLANET AND TRANSFORM
AGRICULTURE ON EARTH**

LENORE NEWMAN
AND EVAN D.G. FRASER

Published by ECW Press
665 Gerrard Street East
Toronto, Ontario, Canada M4M 1Y2
416-694-3348 / info@ecwpress.com

Editor for the Press: Susan Renouf
Copy editor: Jen Albert
Cover design: Michel Vrana

LIBRARY AND ARCHIVES CANADA CATALOGUING
IN PUBLICATION

Title: Dinner on Mars : the technologies that will
feed the red planet and transform agriculture on
Earth / Lenore Newman and Evan D.G. Fraser.

Names: Newman, Lenore, 1973- author. |
Fraser, Evan D. G., author.

Identifiers: Canadiana (print) 20220213518 |
Canadiana (ebook) 20220213674

ISBN 978-1-77041-662-8 (softcover)
ISBN 978-1-77852-029-7 (ePub)
ISBN 978-1-77852-030-3 (PDF)
ISBN 978-1-77852-031-0 (Kindle)

Subjects: LCSH: Food science. | LCSH: Food.

Classification: LCC TP370.5 .N49 2022 | DDC
664—dc23

This book is funded in part by the Government of Canada. *Ce livre est financé en partie par le gouvernement du
Canada.* We also acknowledge the support of the Government of Ontario through the Ontario Book Publishing Tax
Credit, and through Ontario Creates.

ONTARIO CREATES

Canadä

PRINTED AND BOUND IN CANADA   PRINTING: MARQUIS 5 4 3 2 1

To Kat and Christine —
Our partners in crime.

CONTENTS

PART IV: RED DAWN

INTRODUCTION:
The Martian Singularity

THE BOAT PLACE

The men called it the Boat Place.

Captain Francis McClintock arrived at this cursed spot early on the morning of May 30, 1859. He was already uneasy. His small team of men had survived two winters in the high Arctic aboard their steam yacht *Fox*. At first, they supplemented their diet of preserved food with reindeer, narwhal, polar bear, seal, and sea birds, but now they were beyond almost all wild game, surviving only on pemmican and ship's biscuit, fighting scurvy. A few days earlier, they'd eaten some of the sled dogs. McClintock was a seasoned polar explorer, but of King William Island, all he could say was that nothing could exceed its gloom and desolation.

McClintock was in the far Arctic chasing ghosts. Fourteen years earlier, Sir John Franklin led 129 men on a mission to find the Northwest Passage. His ships, the HMS *Erebus* and HMS *Terror*, were fortified bomb-class naval vessels fitted with steam engines and crammed with the latest advances in polar survival equipment.

The journey was supposed to be a triumph of British naval superiority. Instead, Franklin, his ships, and his men vanished, lost in the darkness and cold. It would be the worst disaster in the history of British exploration.

In the years that followed, over forty expeditions were dispatched to hunt for the missing men. Several were funded by Lady Jane Franklin, Sir John's grieving widow. It wasn't until 1854 that explorer John Rae, traveling overland using survival skills he learned from Indigenous North Americans, came upon the truth: Franklin led his men north, his ships were crushed by the ice, and his entire expedition starved.

McClintock was tasked with confirming that this was indeed Franklin's fate,[1] and it was this task that brought him to the lonely horror of the Boat Place. On rolling rocky ground, he found a twenty-eight-foot whaling longboat fitted with sledges for hauling overland. Inside and around the boat was a mishmash of gear: Masses of clothing, pocket watches, hair combs, a beaded purse. Loaded shotguns, sounding lines, crested silverware, and a book of hymns. Silk handkerchiefs, knives, lead sheeting. And in the boat, headless, two sets of skeletal remains, still fully dressed. Between the skeletons, some tea and a considerable cache of chocolate nestled. The sledge was pointed north, towards the false hope of the trapped ships, towards death.

McClintock gathered a few tokens, returned to the *Fox*, and steamed home.

---

1   Lady Franklin refused to believe John Rae as he relied on Inuit testimony instead of crossing to King William Island himself. She blocked his knighthood, and though he discovered the final link in the Northwest Passage during his journey, he did not get credit for the discovery until much later. We will return to John Rae's story later in this book.

"And that, Evan, is what happens when you head out to the great unknown and don't pack enough for lunch. It all comes down to food."

Lenore leaned back in her chair and shivered a little, glancing out her window at the gray rain of Vancouver, Canada. At the other end of the Zoom call sat Evan in Ontario. He was shivering, too, though the warmth of 2020's summer was just starting to push back against a chilly spring.

Separated by half a continent, both of us were cold. And bored. We were each waiting out our own pandemic-flavored Arctic winter. Lenore's apartment felt a little like a sailing ship frozen in ice. Evan had more space, but a larger family. As we chatted, Evan occasionally picked up his laptop and moved to a different room as his teenagers free-ranged through the house.

On that day, we were having a bit of a brainstorm. The two of us had been chatting off and on for about two months. We'd been friends and colleagues for years, but with lockdown, our conversation picked up pace.

In the early days of the COVID-19 lockdown, we mostly moaned about lost travel. Both of us spent much of our pre-COVID lives traveling around the world, studying the global food system. But in March 2020, all those other countries might as well have been on another planet.

And then, one fateful day, we realized there was a place we could go to and study a global food system if we used our imaginations: we could do a thought experiment on what it would take to live on Mars. This struck both of us as a silly idea at first, but as we pondered, this thought experiment morphed into a two-year mission, conducted over Zoom, one cramped claustrophobic room to another.

It was in that moment, in April 2020, when COVID was new, and there was no toilet paper anywhere, the two of us decided we should go to Mars, at least in spirit. And the first question, of course, was what would be for dinner once we got there?[2] While this may seem like an odd question to ask, it is the one in most urgent need of an answer. Nearly two centuries after poor Franklin kissed his wife goodbye, loaded the last casks of fresh water, and sailed over the horizon, humanity is contemplating a journey into an ever-deeper desolation — outer space. And beyond that velvet blackness, Mars.

This book is about what the first Martian community must do to feed itself if it is to avoid the kind of starvation faced by Franklin, Crozier, and countless failed explorers. But despite the Specter of Hunger Past, this story is an optimistic one. As the two of us have gone on this imaginary mission, we've come to believe a Martian community can and will feed itself successfully and, in doing so, will develop technologies that will revolutionize agriculture on Earth.

Seem preposterous? We don't think so. In our day jobs, we are academics. We write serious books, give serious lectures, and advise senior levels of government in Canada and internationally. In all this work, the two of us have devoted our professional energy to developing strategies to sustainably feed the world's growing population. We work on figuring out problems linked to climate change and obesity, how to help people emerge from food insecurity, and the best ways of protecting farmland. Despite all this (or perhaps because of all this), in our opinion, figuring out what the first Mars-dwellers will eat is a topic that may define the future of how we feed ourselves.

At the time of finishing this book, dreams of sending people to Mars are all the rage. Billionaires and space wonks are moving voyages to the Red Planet out from the shadow of fiction and into the

---

2   We could, in fact, have made the actual trip in the time it took for the pandemic
     to pass.

imaginable future. NASA and the European and Canadian Space Agencies are tooling up to establish both lunar and Martian habitations. Some of the world's richest people boast about blasting off to new worlds. .

Perhaps it's the way our horizons have shrunk during the pandemic that's causing this craze for outer space. Perhaps Carl Sagan was right when he wrote, "For all its material advantages, the sedentary life has left us edgy, unfulfilled. Even after 400 generations in villages and cities, we haven't forgotten. The open road still softly calls, like a nearly forgotten song of childhood. We invest far-off places with a certain romance. This appeal, I suspect, has been meticulously crafted by natural selection as an essential element in our survival. Long summers, mild winters, rich harvests, plentiful game — none of them lasts forever. It is beyond our powers to predict the future. Catastrophic events have a way of sneaking up on us, of catching us unaware. Your own life, or your band's, or even your species' might be owed to a restless few — drawn, by a craving they can hardly articulate or understand, to undiscovered lands and new worlds."[3]

Perhaps we've all been sitting too long on our couches and watching too much Netflix.

Whatever the causes, we seem to be on the cusp of a new space race, and the prospect of living on Mars somehow captures the zeitgeist of the moment. But in the back of our minds, we should always remember the frigid winds of the deep Arctic as they play at the tattered sails of a ruined and forgotten boat.

The first human footfall on Mars will be a complete gamechanger for our species, but getting there will also be difficult in a

3   This passage comes from the introduction to Carl Sagan's book *Pale Blue Dot: A Vision of the Human Future in Space*, published in 1994 by Ballantine Books. See page xiv.

way that dwarfs the challenge of trying to force *Erebus* and *Terror* through an ice-clogged Northwest Passage.

People are going to die on Mars. They are going to perish from asphyxiation, and they are going to freeze in temperatures as low as -150°C. They are going to trip and fall in the low gravity and suffer from radiation that, on Earth, is tempered by our lush atmosphere. They will die in spite of the staggering expense to catapult them to the Red Planet. Billions of dollars will be spent on the "Mars Project," money that could be usefully purposed here on Earth, building schools, training doctors, and alleviating poverty. So, at the outset, let's be upfront: Why go? Why bother trying to set up on Mars, which is so far from where we have evolved to live? The risks to the individuals who make the inaugural flights are staggering.

To set the stage for this imaginary voyage, we came up with a little story. As with every part of this book, the story developed over a Zoom video call. In this case, snow was falling in Evan's hometown of Guelph, drifting in bucolic waves. Behind Lenore, Vancouver's lights glimmered in the fog.

"Evan," Lenore began, "I want you to imagine a river . . ."

The river is deep and turbulent, but it divides two prosperous cities. Merchants must brave an expensive, slow, dangerous, and unpredictable ferry crossing to ply their trade.

One day, an engineer stands on the riverbank and thinks how wonderful it would be if a bridge could be built to span the torrent. But the main channel is too wide, and she knows even her best design would collapse under its own weight.

"If only," thinks the engineer, "I could distribute the powerful forces acting on the span to the supporting pillars. Then, I could bridge the gap." Our engineer can imagine the bridge, but she cannot yet build it.

Humanity has solved this problem in several ways — a ferryman punts his boat for a fee; twenty miles downstream is a ford — but the engineer comes up with one of the most elegant of solutions. She creates a design where the stone pillars reach towards each other in a graceful arc. Where the two sides meet, she imagines something completely new: a keystone. The wedge-shaped keystone completes the arch, locks the bridge in place, and distributes the immense weight of the span into the pillars on either side of the bank.

The people who imagine literal and figurative keystone technologies are important. We remember them. They are associated with the great eureka moments of history. The world changes, and the impossible becomes possible.

But keystones on their own are not enough, and just because a thing is possible doesn't mean it is practical. For instance, the engineer might find the cutting of so much stone too slow and difficult. Maybe it is too expensive to transport materials to the building site. Maybe the available mortar is too brittle to bear the weight of the arch. Making a possible bridge a reality requires not only the keystone technology, but that a great many secondary and tertiary problems must also be solved.

The engineer can wait for these secondary problems to be solved through gradual improvement or the development of other keystone technologies, or she can actively work to solve some or all of them herself. Some of these innovations might proceed rapidly. Others might be dead ends as they lack supporting technology. Others still might even work against our engineer, such as the rising cost per hour for stonemasons. But the interesting thing is that the individual problems don't entirely matter; it is the ability to gather up and connect all these supporting technologies and processes that determine whether it is practical to build something entirely new.

Though the subsequent shifts that support the original innovation are more subtle, they are by far the more profound. Suddenly the bridge isn't just possible or practical, it is inevitable. It is built, and ferry service is suspended, and then, up and down the river (and on other rivers of the world), long-span bridges begin to rise. The secondary and tertiary improvements drive exponential improvements in technology and exponential drops in cost. The world experiences a singularity; forever after, there is a time *before* long bridges and a time *after*. From the keystone arch, other innovators move on to develop tied arches, through arches, cable-stays, and suspension bridges. Eventually we are driving across the Golden Gate, and ferry operators all over the world are left scratching their heads, wondering what happened.

"So really we are telling two stories. In one, a failure to anticipate and adapt leads to disaster . . . that's the story of Franklin. But in the other, innovation unleashes waves of positive change . . . our engineer and her bridge."

"Exactly, Evan, and the amazing thing is both are possible when it comes to exploring Mars. Collapse or advance. Failure or survival. And survival makes us stronger."

"And food is the keystone?"

"Absolutely! In truth, it usually is."

Lenore's "Parable of the River" is a fun way of telling the story of innovation writ large. It's the process we will describe in this book as it pertains to humanity's ambitions to set up communities on Mars. And it is also a pretty good fit for describing what happened almost a hundred years ago during the Green Revolution, a period of a few decades where new keystone technologies completely changed food and farming systems and gave us the industrial food system on which most of humanity currently depends.

According to most history books, the grandfather of the Green Revolution is Norman Borlaug who won a Nobel Peace Prize for helping feed the world's growing population. Borlaug is mostly recognized for his work in the 1950s and 60s, when he helped breed high yielding varieties of wheat and rice that were much more productive than previous cultivars. Often, we see Borlaug's plant breeding as a keystone innovation akin to the engineer's arch.

But if we dig into the history of the Green Revolution, we see that plant breeders like Borlaug simply created a keystone piece of technology that helped unify several other technological breakthroughs that had occurred earlier in the twentieth century.

The story goes something like this. Throughout much of history, the amount of water, labor, and plant-available nitrogen limited how much land a farmer could cultivate and how productive that land would be. For ten thousand years, farmers applied their sweat, grit, ingenuity, and determination to overcome these basic limits. Countries developed systems to exploit labor, such as the ancient Sumerians who used slaves to dig huge irrigation canals. Farmers everywhere spread manure and planted leguminous crops to ensure the soil stayed rich in nitrogen. Cows, horses, and even people were yoked to plows. Islands covered in nitrogen-rich bird guano near the equator were taken by military force and harvested to fertilize the fields of Europe and North America.

But starting in the early twentieth century, we began to discover industrial approaches to overcoming these limiting factors. Around 1920, tractors began replacing horses and humans as the primary source of labor in many farming systems. Bigger diesel-powered engines enabled deeper wells and more effective irrigation. A chemist named Fritz Haber and an engineer named Carl Bosch figured out how to produce nitrogen fertilizer through a chemical process.

By around 1940, engineers and scientists had mastered new ways to bring crops more water, nitrogen, and labor than they needed.

But as these old limits to growth were removed, the technologies still didn't unlock huge new vistas of productivity. Nor did they change agriculture for the simple reason that there was a keystone piece of technology still missing. The issue was that when traditional varieties of wheat or rice were grown with abundant nitrogen and water, the plants simply grew too big and too heavy to be able to ripen. These plants literally collapsed under their own weight. Norman Borlaug's contribution was to breed short-stalked varieties that were able to stand erect despite their plentiful harvest. These so-called dwarf varieties were the keystone technology that joined all the other innovations into a moment of singularity.

Suddenly, the metaphoric bridge was built. Ten thousand years of agriculture changed in a generation.

A MATTER OF HORSE SHIT

"But what does this have to do with Mars?" Evan wanted to know after Lenore called him to talk about rivers and keystones and the history of twentieth-century agriculture.

"The point," Lenore replied, "is two-fold. First, figuring out what's for dinner on Mars could be a similar keystone event. There are so many interlocking technologies we need to develop to live on Mars that once we do that, a new way of producing food here on Earth becomes inevitable."

Lenore was pacing, maneuvering around a mound of canned goods and paper products.

"Ok . . ." (Evan was still a bit dubious.)

"The second reason is more pressing. We need to change. Our current system of farming that came out of the Green Revolution has run its course. It's too energy-intensive, too vulnerable to climate change, too polluting, too hard on the land, and, despite using all these tractors, still exploits human and animal labor. This industrial system emerged a hundred years ago, before we worried about climate change and biodiversity. To survive the next 100 years, to keep the Earth from being totally screwed up, humanity needs a new system. A Mars mission might give us another keystone event, a *Mars Singularity*. It's all about the horses!"

"The horses? What do horses have to do with Mars?"

Lenore smiled. She was waiting for that question.

Horses are a useful barometer through which to measure the scale and pace of change during the Green Revolution. In 1920, there were over twenty-five million horses in the U.S., and American farmers devoted forty-five million acres of land to growing oats to feed them. Society was dependent on these beasts as the backbone of agriculture and transportation. Consequently, there was a huge national economy involving blacksmiths, farriers, breeders, and veterinarians.

Yet there were lots of problems linked with horses. In 1898, delegates from around the globe gathered in New York at what is now seen as the world's first urban planning conference. The delegates' job was to find a solution to one of the nineteenth century's key urban problems: what to do with mountainous piles of horse manure. The media of the time suggested that within five decades London's thoroughfares would be drowning in nine-foot piles of horse shit. Meanwhile, in New York, planners predicted that by 1930, manure piles would reach third-storey windows. According to legend, participants were unable to agree to any solution to the problem and the conference ended in acrimony days before it was scheduled to close.

But, of course, the manure apocalypse didn't happen. By 1960, the world had gone through the Green Revolution. Tractors had replaced horses on farms, and trucks were making deliveries all over North America. Irrigation, pesticides, and nitrogen fertilizer were widespread, and the basic economic logic of agriculture had fundamentally shifted. Dwarf varieties of wheat and rice were spreading everywhere, pushing up yields and pushing down hunger. The total amount of food produced on the planet skyrocketed. Today, thanks to these technologies, an incredible 2,800 dietary calories are produced per day for every woman, man, and child on the planet. This means that today, more food is produced per capita than at any time in human history.[4]

And as this transition happened, the number of horses in the U.S. plummeted by more than 85 percent while the acreage of oats fell by around 90 percent. Untold numbers of blacksmiths and horse breeders must have lost their jobs, and the farmers who had grown all those oats had to find new markets or new crops to grow.

But even as worries about drowning in manure subsided, the new industrial system birthed a host of new problems. The tractors, pesticides, fertilizers, and irrigation systems mean that the tools modern farmers need are expensive, so only well-capitalized farmers have been able to survive. All over the world, small-scale farmers have suffered enormously.

Plus, the fertilizers, irrigation, and tractors all require a lot of energy, so this new style of farming yoked our food system to fossil fuels. Together, these new technologies have birthed their own host of problems.

The two of us have written five books and well over 150 articles on this topic.[5] The nutshell of all this work is this: Today, food and

---

4   Yet despite this tremendous bounty, hunger is rising. The reason for this is likely found in the combined factors of climate change, civil unrest, and yawning wealth inequality that has allowed a small number to become super wealthy while the bottom billion fall further and further behind.

5   The keen reader may want to check out Lenore's *Lost Feast* or the book written by Evan along with Ian Mosby and Sarah Rotz, *Uncertain Harvest*.

farming systems are the world's largest contributor to water pollution, the biggest user of fresh water, the largest driver of the loss of
biodiversity, and contribute approximately a third of global greenhouse gas emissions. About 30 percent of the food we produce is
wasted, and the "Western diet," which has spread over much of
the world, is so rich in simple carbs, sugars, and fats that it has
caused obesity and diabetes to become top public health concerns.
The world would need to almost quadruple fruit and vegetable production if everyone were to switch to the kind of diets proposed by
national food guides. In short, our food system is a mess, not fit for
feeding us in the twenty-first century.

If the mid-twentieth century saw the development of keystone technologies and gave rise to today's industrial food system,
the twenty-first century needs another kind of transition to fix a
lot of the problems our current food system has caused. Lenore
thinks that this next transition could come from developing a
self-sustaining community on Mars. Evan started this book a bit
more skeptical, but as we've worked together, he's come to see her
point: to feed a community on Mars we will need to solve many
of the very problems that are bedeviling the food system here
on Earth.

"Ok, here's the key lesson we need to take."

We were getting the hang of communicating over Zoom, and
Lenore was walking Evan through the logic that would ultimately
form this book.

"From the transition of horses to tractors, from cottage farming
to industrial agriculture, we can derive a general theory of innovation we can apply to Mars, and then by applying it to Mars we'll
have everything we need to change things on Earth, OK?"

Evan nodded, not entirely convinced, but willing to play along.

"At first, we can imagine a solution to a problem. It could be any problem but, since we're talking about food, let's imagine a food example, say the need for a country like Canada to be able to produce fresh fruits and vegetables in winter. Scientists, farmers, and policymakers working on that problem will find that some of the things we need to solve that problem are both necessary and impossible — Canada would need to solve *winter* before we can grow tomatoes in January. They are the keystones needing to be found. Many university labs and start-up companies spend decades struggling to find solutions to these sorts of problems.

"Developing human habitation of Mars will be easily one of the most complex problems in human history. It is composed of tens of thousands of supporting problems, branching down in layer upon layer, peppered with missing keystone technologies and processes that are possible but not yet practical. Some of these technologies are already advancing, others are farther away, and still others haven't yet been imagined. But we are moving towards a point at which it will be possible and practical to bridge this widest of rivers.

"We are approaching a point I call the Martian Singularity. Beyond it, inhabiting the Red Planet will become not only doable but inevitable. And in doing so, we'll also have developed a critical mass of intersecting technologies that will have propelled those of us left on Earth into a new paradigm of food production and agriculture."

"OK," said Evan. "Let's plan dinner on Mars."

A continent apart, the two of us got to work.

PART I:

RED

HORIZONS

CHAPTER 1:
Arrival

W e are big science fiction fans. We love the escapism, but we also love how good sci-fi provides a way of exploring how the future may unfold. The best science fiction offers a thought experiment about how humans might live under different conditions and, in doing so, gives us fresh ideas about our problems today. Despite being dated and problematic, we think world-builders like Robert Heinlein and Isaac Asimov are a pleasure to read. And we both agree that the early 2000s reboot of *Battlestar Galactica* is amongst the best TV ever made (but don't get us started on the ending). The only reason either of us has a Prime membership is for *The Expanse*.

So, indulge us, for a few moments, while we try to paint a picture for you of what a Martian city might be like.

Imagine that today's date is in the 2080s and *you* are part of the third or fourth wave of pioneers to the Red Planet. Perhaps you are a teacher brought in to educate the first generation of Martian-born children. Or maybe you are an entertainer or an artist whose job it is to inject some culture now that the basic building blocks of life have been established.

Also, imagine that thanks to climate change, life on Earth has become challenging for most people. Things haven't gotten better since the 1920s, our leaders did not tackle issues with courage, and so overpopulation, inequality, poverty, and water scarcity have made major metropolitan areas extremely unpleasant. The Earth, for many people, has become hot, dirty, and crowded.

But in this future, an alternative has opened up: a permanent community has been established on Mars, and people are lining up for the opportunity to go. And it's not just engineers, astronauts, and scientists who are going. The UN is now taking applicants for pioneers who come from many walks of life. The explorers and scientists went first, and before you signed up for the selection lottery, waves of robots had been descending on the Red Planet, laying the groundwork and infrastructure for humans.

The first mechanical pioneers arrived early in the twenty-first century. These machines simply scouted the environment, analyzing samples of the Martian regolith, which is mostly sand, dust, and gravel on top of bedrock. They beamed back pictures of a fantastical landscape that ignited the imaginations of folks on Earth.

Those robots were backed up with satellites put into orbit around the Red Planet. Together, the rovers and the satellites did a pretty good job identifying the most likely places for human habitation. Initially, there seemed to be a handful of possible locations where it might be conceivable to establish a human base.

We can imagine the candidates. The first location was a nondescript crater close to the north pole located in Vastitas Borealis, a gargantuan flat plain that stretches across much of Mars's northern hemisphere and is dotted with similar impact craters (the thin Martian atmosphere means that random space debris hits the surface of Mars much more often than on Earth). The crater in question is located at approximately 70.5° north and 103° east and has a diameter of approximately thirty-five kilometers and a depth of two kilometers. This crater is entirely uninteresting except for one key feature: a large patch of frozen water. Initially discovered by the European Space Agency's *Mars Express* spacecraft in 2005, the benefits of locating a permanent base near a source of water speak for themselves. One of the key things any permanent community needs is access to H_2O, for sustaining human life and as a source of fuel (water molecules can be split to create oxygen and hydrogen, which can be used for fuel cells). It is also necessary to grow food. A crater full of water is just as nice on Mars as it is on Earth.

A second likely location is farther south, in the mid-latitudes. There, it would be possible to tunnel into the sides of the cliff in the Tempe Mense region at 40° north, 70° west. This site was identified as a possible location for a Martian city by the architectural studio Abiboo in 2021.[1] Tempe Mense could provide shelter from the harmful radiation that bombards the surface of Mars. Martians will worry about radiation a lot.

Regardless of the location, the cost of setting up a community proved too great for any single country or corporate billionaire alone. But the desire to see a human community on a different planet was too strong to resist. So, by the mid-2030s, dreaming of Mars opened an unprecedentedly collaborative, and entirely new, chapter of human history. Initially, the diplomats, titans of industry, and scientists debated the right location, and after much wrangling, eventually settled on the cliff face of Tempe Mense. While everyone agreed that this area didn't have the abundance of water of the north or south poles, detailed reconnaissance had discovered ground ice (somewhat like permafrost on Earth) in abundance in most mid-latitude areas. This ground ice would do in a pinch, and most agreed that Tempe Mense provided the would-be Martians a sort of Goldilocks zone.

---

1   This actually happened, and here is a link to the concept drawings, which Evan and Lenore think are amazing: https://abiboo.com/projects/nuwa/

It is slightly warmer than the poles, has a reasonable supply of water in the nearby permafrost, and the large stable cliff face could be tunneled into to provide the basis for human habitation protected from harmful radiation.

After this decision had been made, the next wave of robots started to arrive. Diggers that scooped up the ice-filled regolith, a small autonomous nuclear reactor for power. Then came a flock of robotic solar panels that landed on the top of the cliff and unfurled like butterfly wings. Foundries were shipped at great expense for turning local rocks and minerals into steel and cement. An entire battalion of orbital space mirrors and balloon-mounted solar concentrators focused the sun's energies onto the solar panels, helping to augment the nuclear power.

Next came the 3D printers and a combination of specialty manufactured components, such as computer chips, from Earth. These were used to bolster what the space wonks call ISRU (for in situ resource utilization), and this gave rise to the first generation of large robotic machines that began tunneling into the cliff face, constructing first modest and then more elaborate spaces that would become human habitation. UV-filtered glass, carefully crafted mirrors, and lots of locally constructed fiberoptics allowed natural light — but not harmful radiation — to penetrate inside the caves.

We can imagine the setbacks the Martian missions will inevitably face. About midway through this process, a couple of machines went haywire, and a human maintenance and repair crew had to be dispatched to troubleshoot. Within ten years, most of these first-timers were dead from cancers caused by radiation exposure. Then there was a fire in one of the early habitation zones that killed five engineers. And there were inevitable cost overruns as well as a major diplomatic crisis sparked by what the new community should be named. Frantic negotiations by the world's corporate and political elite resulted in a number of proposals: Jamestown, New Dubai, and either BezosVille or Musklandia[2] were all serious contenders. In the end, everyone simply ended up calling the place BaseTown, and the vernacular stuck.

So here *you* are, sixty years after those initial American, European, Indian, and Chinese rovers sent back pictures of a red landscape, and you are landing in BaseTown to begin a new life.

The community itself has grown to about ten thousand people and is expected to reach one hundred thousand over the next generation. BaseTown still depends on regular shipments of materials and manufactured goods (to say nothing of luxuries such as coffee or chocolate) from Earth. But the business case to this whole experiment is starting to look plausible. The technologies developed to build BaseTown have enabled robotic mining operations in the asteroid belt and act as a staging ground for exploration of the outer solar system, giving the community an economic impetus.

More importantly, however, the technologies developed to sustain life on Mars have started to revolutionize life back on Earth. Cyanobacteria bred to turn the nitrogen and carbon dioxide in Mars's atmosphere into organic molecules have been exported back to Earth. Back home, these cyanobacteria now mop up the carbon dioxide left in the Earth's atmosphere by the industrial age and help rehabilitate degraded farmland. The way BaseTown synthesizes proteins in giant bioreactors has also been exported back to Earth to keep people healthy now that climate change has limited animal agriculture.

However, you don't see these bacterial bioreactors when you take the transit pod from the orbital landing pad to BaseTown's cliff-face entrance. The bioreactors don't make much of an impression

---

2 We can also imagine that Richard Branson's heirs wanted the place called something like Virgin Territory, but we would hope that this proposal would have been rejected outright.

on the landscape. Really, they are nothing more than giant underground tanks fed with solar energy and carbon dioxide. Sunlight is brought down using fiber-optic cables, so most of BaseTown's inhabitants are only faintly aware of this part of the infrastructure.

What catches your attention — what catches everyone's attention — is the cliff face that has become a honeycomb of reinforced windows and skylights, carefully designed to withstand the possibility of a meteor strike, tinted to filter out harmful radiation, but engineered to allow in the greatest possible amount of light. When you step out of the transit shuttle and into BaseTown's main mezzanine, you are greeted with a surprisingly open and well-lit space that is full of greenery. One of BaseTown's abiding design principles is to ensure that every photon of solar energy that enters BaseTown is put to multiple uses. Plants are encouraged to grow in all sunlit places. And each of these plants sucks up extra carbon dioxide, emits oxygen, purifies water, and turns organic molecules into food. Plus, the plants are nice to look at: it's hard and dangerous to take a trek out on the surface, and most people spend their time on Mars indoors. Plants outnumber people by a thousand to one.

These plants have been bred to deal with the low-gravity and lower-pressure environment and all are nestled in a hydroponic solution, but your first impression of BaseTown is that every resident must be a master gardener.

It's been a long journey from Earth, and for months you have been eating nothing but a steady diet of freeze-dried and hyper-preserved food. So, you are eager to taste Martian cuisine. Identifying a small leafy food court, you use your new identification card and order yourself salmon sashimi, toasted seaweed, a garden salad, and then, glancing at the beverages, you opt for a milkshake.

On the menu, all these items are listed as locally sourced, and when they arrive on a conveyor, you are pleasantly impressed.

The sashimi is really a cell culture that has been 3D printed to look as if it came from salmon; as you bite down on the pleasant texture, you wonder whether this technology might have prevented the Earth's salmon from being driven to near-extinction by overfishing.

The seaweed tastes exactly like the seaweed on Earth, as does the green salad, except that this green salad is much fresher than what you are used to as it has been harvested from hydroponic beds earlier today. And as for the milkshake, the dairy proteins are the product of a specially designed fermentation process, but you have no idea whether this milkshake tastes as if it came from actual cow's milk because you've never tasted "real" dairy before. Over the past two generations, the number of cows on Earth plummeted, and today it is rare to eat real meat or dairy, and many people choose not to. Regardless, this Martian milkshake is both refreshing and rich.

After you finish this first meal on Mars, you carefully place your utensils and dishes in a sterilization chamber and meticulously scrape every scrap of organic matter off your plate and into a sophisticated composting infrastructure. The UN's training for space pioneers was very clear on this point. On Mars, there can be no such thing as waste. Resources are scarce, and everything must be carefully stewarded. Again, perhaps this is another lesson that we should have learned ages ago on Earth.

You look around at your new home — your fresh start — and think, *This might not be so bad. Maybe folks back home have something to learn from the system we've built up here?*

CHAPTER 2:
Foundation

THE JACOB TWO-TWO CHALLENGE

"The first thing I think we need to do," said Evan to Lenore one sunny afternoon as the draft of this book was coming into shape, "is to address the obvious. Someone is bound to tell us that it will be simpler to send food to Mars than trying to figure out how to produce it on a planet that, as far as we know, has never hosted life. Maybe rockets, space elevators, gravitational slingshots, or even giant catapults can throw the basic ingredients to support BaseTown more easily and cheaply than trying to engineer an entire ecosystem in an environment that is frigidly cold, has very little solar energy, no liquid water, and zero topsoil?"

Lenore pondered this. She needed to crunch some numbers.

A couple of days later, the two of us reconvened so Lenore could lay out a crisply reasoned set of arguments. But she began the discussion in an unusual place . . .

"Evan, do you remember that seventies kids' book *Jacob Two-Two Meets the Hooded Fang*?[1] The story starts when Jacob's mother sends him down to a shop to buy veggies. Well, what if Jacob was a Martian, and the greengrocer was dependent on shipments from Earth? Even ignoring the months and months it would take for the tomatoes to arrive, the cost of getting those two pounds of firm red tomatoes to Mars would be crazy high.

"I mean, NASA estimates that it costs $10,000 to put one pound of payload into Earth orbit, so if those tomatoes came from Earth, just getting them into space, let alone the cost of transporting them to Mars, and then the reentry onto the Red Planet, means that a mayonnaise and tomato sandwich at a takeout diner in our hypothetical BaseTown would cost at least a couple of months' salary for an average NASA astronaut."[2]

With those timelines and economics, Mars is too far for takeout. Self-sufficiency, food sovereignty, and self-reliance will be the order of the day. And this means we need to design whole ecosystems from the bottom up. Martians will have to eat locally.

To imagine how to do this, we need to understand the foundations of food systems here on Earth. At the base are what ecologists call autotrophic organisms — the primary producers that create their own food. Plants are the most widely known key autotrophs because photosynthesis allows them to turn sunlight into oxygen and sugar. Plants form the foundation of most terrestrial food chains and are the stuff that herbivores eat (before the herbivores are consumed by carnivores. After living matter dies, it is returned

---

1   It's a strange story, really. The mom sends a four-year-old to the store.

2   According to NASA's webpage in 2021, the starting salary for an astronaut is about $65,000 U.S. per year.

to the soil by detritivores, closing the loop). This means any Martian ecosystem needs to start with autotrophs. But plants don't just need sunlight and carbon dioxide, they also need water and a thin layer of topsoil rich in organic matter and vibrant with microbial activity. Plants are not actually autonomous but are part of a complex set of ecological relationships, each of which involves many different species and taxa working together.

Take nitrogen for instance. On Earth, nitrogen is abundant in the atmosphere where it takes the form of the N_2 molecule. But the atoms of nitrogen in N_2 are bonded together so tightly they are not available for plants to use. Since nitrogen is a basic building block of protein, this lack of plant-available nitrogen is a problem that limited the productivity of most preindustrial farming systems. However, some species of plants (especially legumes such as peas, beans, or clovers) have a symbiotic relationship with soil-bound microbes that can take the nitrogen from the air and "fix" it into forms of nitrogen that plants can utilize. In exchange, the nitrogen-fixing microbes feed on the plants, and both win. This is why traditional farming systems all over the world rotate legumes with grain crops as a way of keeping the soil fertile and yields high.

The way that legumes work with microbes to fix nitrogen is just one of countless other ways in which soil, microbes, and plants work together. But of course, on Mars, no such evolutionary legacy exists. On Mars, we will be forced to start from scratch.

This is a good news/bad news story. The good news is that we have some of the key building blocks of life on Mars already. The atmosphere is rich in carbon dioxide and N_2. There are plenty of potential plant nutrients such as phosphorus in the regolith. Crucially, the regolith also contains frozen water that could be melted and utilized. Water might also be mined from the polar regions where at least one frozen lake has been found. And there is some solar energy.

But the obstacles are huge. In general, Mars is cold. Really cold, minus-sixty degrees Celsius on average (but ranging from minus 150 degrees Celsius at the poles in the winter to twenty degrees Celsius on the equator during summer). Being a smaller planet, the gravity is only about one-third of Earth's, which means the atmosphere on Earth is about one hundred times denser than the atmosphere on Mars. The Red Planet is much farther away from the sun, so it only receives about one-half of the solar energy that Earth does, but because of its very thin atmosphere, damaging radiation is common. Finally, Mars's regolith also contains compounds called perchlorates (e.g., ClO_4-) that are toxic to humans and plants.

The challenges of designing functioning ecosystems out of this are staggering. Nevertheless, it's possible to break this into a series of steps, each of which is amenable to some sort of solution.

First, we need water, but we know that there is plenty of water in the polar areas that could be mined as well as more modest amounts frozen in the Martian regolith. So, this water could be harvested and melted to support plant growth. Martians will have to be very careful not to waste water, but this won't be an insurmountable issue.

Then there's energy. To get started on Mars, we would almost certainly need some sort of small nuclear generator that would have to be imported from Earth. But BaseTown would also need to harness as much solar energy as possible. Conventional solar panels would be needed to generate additional electricity (though would have to be fitted with a self-cleaning surface in case they get covered during a Martian dust storm), but Mars will also need solar concentrators, filters, and collectors (essentially mirrors, lenses, and fiber optics) to focus the limited amount of sunlight that reaches that far out in the solar system and filter out the harmful parts of the spectrum. Such collectors and concentrators would need to be deployed in a number of ways: on the ground, on satellites in orbit, as well as mounted on weather balloons.

But even with these issues solved, we are still missing at least one key ingredient that the ecosystems of BaseTown will need. Plants do not live on water and sunlight alone, so our Martian farms will need a rich supply of organic molecules and beneficial bacteria on which to thrive. This is where we need to look beyond plants to other primary producers as the autotrophic basis for the food system. We need cyanobacteria. On Mars, the food pyramid will have a different base layer composed of tiny, hardworking organisms that can form nutritional organic substances from simple inorganic components.

IN PRAISE OF BLUE-GREEN ALGAE

Amazing things can be found in bogs. Lenore spent many a happy childhood hour studying the frogs and lizards in nearby swampy patches of rainforest, returning to her house covered in a mind-boggling variety of mud and slime. The goo floating on stagnant water is unloved, but there is chemical magic in those stinking waters. Part of the magic comes from cyanobacteria, a bacterial organism capable of photosynthesis. The Greek origins of the word share similarities to the name for the blue-green color cyan, and although modern botanists don't like referring to cyanobacteria as algae (technically, algae are a different kind of organism altogether), many people know of cyanobacteria simply as blue-green algae.

On Earth, these organisms get a bad rap. In the summertime, especially around lakes in farming regions, over-fertilized fields and livestock operations often cause nutrients to run into waterways. Under these conditions, cyanobacteria can bloom and spread in giant mats of floating scum, and they can sometimes produce extremely poisonous toxins called cyanotoxins. When that happens, things become downright dangerous. But even when the blooms don't produce cyanotoxins, when cyanobacteria die, the bodies of these microorganisms simply slip beneath the surface of the water and decompose, absorbing so much oxygen from the lake that fish and marine life suffocate. Lake Erie, one of North America's Great Lakes, nearby Evan's grandfather's old farm, suffers regularly from these blooms. As such, Evan finds it ironic that we may be turning to cyanobacteria as the primary way of establishing life on Mars.

No matter how tarnished cyanobacteria's reputation is on Earth, this microorganism has several key properties that make it an ideal autotroph for the Red Planet. We know this in part due to research conducted at the University of Bremen, in Germany, where researchers grew cyanobacteria in a series of stainless-steel tanks that were fed a mixture of gases that simulated the Martian atmosphere.[3] The scientists also subjected the cyanobacteria to an atmospheric pressure approximately ten times lower than that of Earth. Part of the experiment involved feeding the cyanobacteria with minerals and nutrients designed to match those that are common in the Martian regolith. Despite all these constraints, the cyanobacteria grew reasonably well, fixing nitrogen, producing oxygen, and leaving nutrient-dense organic matter at the end.[4] The experiment concluded that it should

---

3   Verseux, C., Heinicke, C., Ramalho, T.P., Determann, J., Duckhorn, M., Smagin, M., & Avila, M. (2021). A low-pressure, N_2/CO_2 atmosphere is suitable for cyanobacterium-based life-support systems on Mars. *Frontiers in Microbiology, 12*, 67.

4   While this experiment was done at more atmospheric pressure than you find on Mars (Mars's atmospheric pressure is about 1 percent that of Earth's whereas the experiment showed you could grow cyanobacteria at 10 percent Earth pressure), the fact that the cyanobacteria grew well at 1/10th of Earth's atmospheric pressure means that when the inhabitants of BaseTown build cyanobacteria growing facilities on the Red Planet, these can be made of relatively lightweight material as they don't have to withstand too great a pressure differential.

be possible to put cyanobacteria at the base of the Martian community and use it to turn locally found Martian ingredients into a food system that could function without regular supply runs from Earth.

But even if we get cyanobacteria to digest Martian minerals and gases and produce oxygen, nitrogen, and organics, there are still other challenges.

One problem we mentioned earlier is that Martian "soils" contain a lot of the highly toxic compound perchlorate. During some of the early NASA missions to Mars, the rovers found perchlorate in the regolith. This, and later observations made during subsequent missions, leads NASA to estimate that there is about half a gram of perchlorate per liter of regolith. That's enough to cause lung cancer in humans and really inhibit plant growth.

Top-quality air filters could solve the problem of keeping the perchlorates out of BaseTown's breathable air. But the perchlorates will still need to be removed from the regolith before it is fed to the cyanobacteria if it is to be prevented from entering the food system. But here, too, there are scientists working on a microbial solution.

Research done at several universities, including the University of Rome Tor Vergata, are using something called synthetic biology to create biological life-support systems that use a range of microorganisms (including cyanobacteria) to filter out the harmful elements of perchlorate.[5] As perchlorates contain oxygen, it seems that harnessing microorganisms to digest these highly toxic substances can clean the Martian regolith and generate breathable oxygen at the same time. The experiments are still preliminary, but NASA is funding this endeavor in the hope that breakthroughs will allow genetically

---

5   Billi, D., Fernandez, B.G., Fagliarone, C., Chiavarini, S., & Rothschild, L.J. (2021). Exploiting a perchlorate-tolerant desert cyanobacterium to support bacterial growth for in situ resource utilization on Mars. *International Journal of Astrobiology*, 20(1), 29–35.

engineered microorganisms to digest Martian materials. This would form the basis of designer ecosystems for the Red Planet.

This requires several steps.

First, robotic excavators would harvest the regolith. Then it would be warmed up with solar collectors to melt and extract the water. Third, robots would put the regolith in a series of tanks kept warm enough to sustain different communities of microorganisms. These microorganisms would use atmospheric carbon dioxide and nitrogen gas to digest the regolith and break the perchlorates into harmless compounds. In doing so, the microorganisms, including cyanobacteria, would fix the nitrogen, create oxygen, and produce organic matter.

But that wouldn't be it. Next, the community would need to figure out what to do with the organic matter. An obvious approach would be to take the organic matter and turn it into things humans eat — maybe the chefs on BaseTown will cook up some good algae recipes by mixing it with fats and seasonings to create kibble as in *The Expanse* TV series or even produce plain old burger-like patties. Evan, to be honest, is hoping this isn't the main outcome because it would be hard to imagine folks in BaseTown eating algae paddies seven days a week. However, some species of algae produce a decent cooking oil that you can buy today in some health food stores, so algae are going to play a big role in the food systems of Mars.

Another approach would be to use the organic matter as fertilizer and grow plants in greenhouses and vertical farms, producing all manner of crops. This approach allows for a much more varied diet, and dovetails with another option, using the sugars and other useful molecular matter produced by the cyanobacteria as the feedstock for another set of microorganisms, such as yeasts, that would be engineered to digest the dead algae to produce synthetic proteins such as cellular beef, chicken, and dairy. We will come to this strategy in detail later.

Evan was in an autotroph frenzy as he explained this system to Lenore, his hand gestures moving in and out of the video call. "Once you've built the big tanks on Mars for growing the microorganisms and generate enough energy and heat through the solar collectors to run the thing, then all you should need is a few cell cultures from Earth to get the system going. From there, you'd have everything you should need to create a Fully Autonomous, Self-Sufficient Biological Life-Support System. I think we should call it FASStBLiSS."

Lenore tried to absorb this: Are algae and bacteria the key to living on Mars?

UNDERSTANDING THE TOP THREE INCHES

"OK," came Lenore's cautious reply, "compelling points, elegantly simple, but two things. One — we have a 'life-support system' on Earth called the environment, which is hugely complex with millions and millions of species that create the world's ecosystems. Together, these ecosystems produce the air we breathe, the food we eat, and purify the water we drink. We have evolved with it over countless millennia. FASStBLiSS sounds great, but isn't it a tad simplistic?

"Second, someone at some point is going to ask, 'What's the benefit of developing all this Martian stuff for folks here on Earth?' Unless you're Elon Musk, Richard Branson, or Jeff Bezos, it's the taxpayer who will be footing the bill for the trip to Mars. And the only way the space agencies are going to get the funding for this sort of endeavor is if there are benefits back here on Earth."

Lenore's two critiques neatly summarize some of the key arguments against the Mars mission — namely that it is hubris to think we can engineer life-support systems that took millions of years to evolve on Earth, and that if we (as a society) end up spending billions of dollars to send a handful of folks off-planet, then we really need to ensure that in doing so we also make life better for folks here on Earth.

The town of Elora, Ontario, is about as far from a Martian mission as can be imagined. Wandering in its town center, or under the mature hardwoods that grace the century-old residential avenues, Elora gives visitors a glimpse into the once vibrant rural small-town life that flourished before big box stores, sixteen-lane highways, and strip malls. While it's true life in the "golden age" of small-town North America had its own share of problems, it was certainly pretty. The red and yellow brick houses are tidy and well-appointed; the downtown bustles with a hint of old-world charm that borders on the twee. It is a set piece for an early twentieth-century period drama and is the sort of place where people worried about the future seek refuge.

This is ironic considering that just to the south of town sits a world-leading space-age soil science facility called a lysimeter.

From the road, the lysimeter looks like a nondescript corn-field with a trailer parked off to the side.[6] But as you approach, it's possible to make out eighteen large, circular, almost futuristic, shapes on the ground, clustered into groups of six. Each of these circles is about one meter in diameter and is the top of an enormous one-and-a-half meter soil core, originally collected from two different regions and carefully shipped to this field. Each core has been delicately placed in a hole such that the top of the

---

6   Lenore adores both cornfields and horror movies that take place in cornfields, particularly in the autumn when the corn is dry and brittle.

core is level with the rest of the field but sits, deep underground, on a weigh scale so sensitive that the computers collecting the data must be calibrated to compensate for the effect of the wind blowing across the field.

Each soil core is also ringed with a fine lattice of sensors that measure the way water and nutrients flow through the soil, and the whole facility is controlled from three underground bunkers accessed via a series of hatches built into the ground. Climbing down the ladder into these chambers feels like going into the command control center for a nuclear submarine — or a Martian habitat.

Professor Claudia Wagner-Riddle, a soil science professor at the Ontario Agricultural College in Canada, runs the facility with an international team of co-investigators and uses it to explore how agriculture affects soil health and greenhouse gas emissions.

"Basically," she explained as she showed Evan around the facility, "we plant crops on the top of the eighteen lysimeters, just as a farmer does, and use the sensors and computers to measure the impacts of growing these crops." Doing this allows her to answer questions like, "Do fields with diverse crops emit more greenhouse gases than fields where only one crop is planted?" or "What is the best way of building up soil organic matter to help trap water and reduce the impact of climate change?"

Professor Wagner-Riddle does this science to better understand how the soil works, both so that we can protect it and also so that we can learn how to harness some of its amazing properties.

But before she and Evan jumped into the science, Claudia described how her interest in soils and climate change emerged during her years growing up in Brazil. "As a child," she told Evan, "I watched fields full of perennial crops such as coffee be replaced by annual crops such as soybeans and corn that caused tremendous soil erosion."

Claudia later spent time working as a summer student on agricultural systems where she witnessed firsthand how no-till farming (where farmers plant crops without physically disturbing the soil), can reduce erosion and limit greenhouse gas emissions.

"My family on my mother's side were German immigrants who came to Brazil in the 1930s. At that time, the region I grew up in was mostly forest but shifted soon to coffee plantations. Then, in the 1970s, when I was a teenager, there was a big frost and a lot of the coffee plantations died. This is when annuals like soy started to be planted, and this drove a huge amount of soil erosion."

This early start propelled her to an international career that landed at the University of Guelph where she is the energy behind the lysimeter network. She uses the lysimeters to explore how farm management — such as rotating crops from one year to the next, planting cover crops over the winter, and other basic farming practices — affects the health of our soils.

"This facility looks high-tech," she said, "but really, we're investigating relatively simple ways to keep soils healthy. Things like adding extra crops to what a farmer plants, as this can help build up the amount of organic matter in the soil."

The point she makes is that farmers here on Earth can do a lot to protect their soils, and when they do, a whole host of good things cascade out.

"You have to remember that soils with lots of organic matter trap carbon and act like sponges, absorbing both water and nutrients when they are plentiful and storing them for when they are needed. Crops planted on soils high in organic matter need less fertilizer. And building up soil's organic matter essentially takes carbon dioxide out of the atmosphere, helping mitigate climate change.

"Properly done, well-managed soils are the basis of what's called 'regenerative agriculture,' which basically means turning agriculture from being one of the key culprits that drive climate change to a solution for one of the world's most pressing problems. And some of the best things a farmer can do is plant things like cover crops that don't get harvested, try to disturb the soil as little as possible, and embrace diversity within the field."

"OK, so let's talk about scale for a minute then," Evan interjected. "Do the results of your experiments mean that different scales of farm operation are better or worse for the environment? I mean, a lot of folks say that big farms result in monocultures, and we need to be focusing on small farms."

"I'm not sure," came Claudia's thoughtful reply. "I think that farming at all scales can be good or bad — it's really how it's done. If a big farm uses good crop rotations and matches its operation to local soils, then even very large farms can be good for the environment. The reverse can also be true and small-scale farms that feed local markets can — if the farmers don't pay attention to the soil — be net emitters of greenhouse gases."

"But how can farmers actually do this?" Evan wanted to know. "And, do you think learning about Earth-bound soils will help us if we ever reach for the stars?"

Claudia answered both questions immediately.

"First," she said, "by understanding how soils work, which includes both their chemical composition as well as the microbes that live in the soils, we can better manage our natural resources, and this will help not only keep us alive on Earth but also give us tools that may work in other places.

"Second," she continued, "one of the things we're studying is how nitrogen works in the soil. This may lead to some breakthroughs that will allow us to engineer bacteria so that non-leguminous crops are able to start fixing their own nitrogen. Imagine, never needing to apply fertilizer to corn or wheat because we engineer bacteria to work with those crops and pull it directly from the air? That would be huge!"

And she's right.

Today, to keep crops productive, farmers add a lot of nitrogen to the soil. Producing and applying this nitrogen takes a lot of greenhouse gas–emitting energy, but because the amount of nitrogen a plant will use on a given day depends on the weather, it's hard to predict how much and when to apply fertilizer. As a result, many of the nutrients applied to crops don't get applied at the right time or in the right amounts and end up running off fields and into waterways, making agriculture the world's number one source of water pollution.[7] But Claudia's research on how the soil traps and stores nutrients may give us ways of being much more efficient. If this sort of insight could be reproduced in a lab, it would not only reduce the impact of farming on Earth but give us the tools we will need to turn the sterile Martian regolith into something that would support organic life.

Lenore and Evan adore this sort of technology. But they know it will take decades before we see sustainable communities off this planet. Unfortunately, the trends here on Earth — the loss of biodiversity, the water pollution, climate change — could destroy the terrestrial ecosystems we depend on long before we have a shot at the stars. In many ways, the future seems just so bleak. Is it possible to pivot Earth's food system in time to avoid ecological disaster? One possible piece of that puzzle once again focuses on the very small. In the case of microbes, we might perfect our understanding on Earth long before we go to Mars.

---

7   And plastic pollution — a huge amount of which comes from food packaging — is a second massive source of water pollution, so we're in a situation today where the nutrients running off our farms, and the plastic wrappers consumers throw away, constitute a massive blow to aquatic and marine ecosystems everywhere.

FARMING MICROBES

Harnessing cyanobacteria for Mars and Claudia Wagner-Riddle's research to protect the topsoil of Earth come together in one important scientific concept: the microbiome.

What we call the microbiome is the community of all the various microscopic critters (bacteria, fungi, protozoa, and even viruses) that live in a particular environment. And, as scientists have begun to show over the past few decades, the microbiome is important. For instance, the human microbiome, which has been catapulted to fame in recent years, refers to those microscopic organisms that live in and around our bodies. The microbiome of our intestinal tract has received a lot of press because having a distressed microbiome is linked to a range of health problems. One leading hypothesis for this is that the overuse of broad-spectrum antibiotics to fight minor infections, combined with the Western diet's proliferation of sugars, simple carbohydrates, salt, and saturated fat, may have changed the composition of these microscopic communities. This may be behind everything from gluten sensitivity, to anxiety, and depression, and even to asthma in children. Autoimmune diseases (which include diabetes and rheumatoid arthritis) are also linked to a dysfunctional microbiome. A consensus is emerging that a healthy gut microbiome is critical to good health.

Of course, all creatures have a microbiome. For instance, as discussed above, the microbiome that exists between the roots of leguminous plants and the soil is what allows peas and clovers to fix nitrogen out of the atmosphere. And using cyanobacteria and other microorganisms to digest Martian regolith illustrates how we can harness the microbiomes we all already depend on in new ways. What's becoming clear to soil scientists is that nitrogen-fixing bacteria are just the tip of the iceberg.

Healthy soil microbiomes are vital to maintaining healthy soils, and healthy soils enable plants to grow, keep crops alive during drought, and help fight off pests. Just as a healthy human microbiome protects people against a range of ailments, a healthy soil microbiome is key to a healthy farming system.

Work to harness insights about the soil's microbiome is underway in the form of a new type of agricultural input called biofertilizers. Defined as microbial inoculants with both living and dormant cells, biofertilizers contain microscopic organisms that make key plant nutrients (such as nitrogen and phosphorus) available to their plant hosts. In other words, *bio*fertilizers don't provide nutrients to the plants themselves but rather they create microbial pathways that increase our crops' ability to access nutrients already in the environment. Here on Earth, farmers are already using cyanobacteria as a bio-fertilizer that lets crops capture atmospheric nitrogen. Biofertilizers do not require much energy to produce, and they help build up soil health.

Just as Western diets have degraded our guts' microbiomes, making many of us susceptible to a range of ailments, so, too, has industrial agriculture impoverished the soil's microbiome. Current industrial agriculture depends on simple cropping systems dominated by a handful of major commodities that rely on pesticides and herbicides to protect against pests and fertilize with nitrogen, phosphorus, and potassium to keep yields high. While these tools have been incredibly successful in boosting yields (the average farmer today gets over 200 percent more harvest per acre than they would have in the 1970s), this industrial farming has deteriorated soil health.

A mission to Mars could supercharge this whole field of study by sending scientists like Wagner-Riddle on a quest to identify how different microbes benefit farming on Earth. This would give us the chance to create a catalogue of the microorganisms we would need to produce food off-planet. Such a scientific quest would also increase our understanding of how we benefit from the Earth's microbiome and then we could use this as the basis for creating designer ecosystems on Mars. In other words, a trip to Mars would force us to better understand how the Earth's microbes work, and this would give us an edge on Mars.

But the real beneficiaries would be the farmers (and consumers) on Earth who are going to need the same insights to survive twenty-first-century climate change. Most of us will have to rely on Earth's ecosystems for the duration of our lives. And one legacy of our industrial society is that we've changed our climate and our ecosystems so radically that the twentieth-century ways of producing food simply need to change. A Mars mission would help fund a new generation of agricultural research.

Consider the Oloton corn of Oaxaca, cultivated by the Mixe people in the high mountains. A fascinating illustration of the potential of this approach, it is a giant, slow-maturing variety of corn that grows well despite being planted in infertile, nitrogen-poor soils. This corn reaches sixteen feet in height but is also unusual as some of its roots sit above the soil where they are exposed to the air. These tuberous aerial roots are also covered in a wet thick mucus. The plant came to the attention of agricultural scientists in the 1980s when Howard-Yana Shapiro explored the last isolated mountain villages in the region. Other scientists followed, but it would be four decades before a team from University of California, Davis understood the science of this indigenous corn variety.

The team, led by Dr. Alan Bennett, used DNA sequencing tools to determine what species in the microbiome of the root mucus could fix nitrogen. Once these were isolated, the next step was to try to develop inoculants for plants in other regions. Doing all these forensics took over a decade, and this project also involved careful collaboration with the community to ensure locals get a share of any commercial benefits unlocked by this genetic legacy.

The technologies to do this sort of DNA sequencing have exploded in recent years aided by more powerful computers and artificial intelligence. A mission to Mars would galvanize a generation of researchers here on Earth to better understand the way we all depend on the microbiome. Farmers on both planets will benefit.

This was the first brick in Lenore and Evan's Martian food pyramid. A carefully engineered microbiome would underpin all life on Mars.

CHAPTER 3:
Small Is Beautiful

THE WEIRD WORLD OF NANO

Expectations were high for the summer of 2021. COVID numbers were down, talk of near normal life abounded. Lenore was beginning to look to the border with a hungry longing, spending long hours on websites like Rome2rio and Expedia, planning imaginary trips. That summer had different ideas, and the two of us mostly stayed bunkered down, exploring a different sort of frontier, a frontier even smaller than that of the autotrophs.

Evan had decamped with his daughters to the old family cottage on Georgian Bay and reunited with his parents whom he'd not seen in a year. In Vancouver, Lenore was baking under days that went up to forty degrees Celsius as a climate-change-induced heat dome was smashing records, killing people, sparking wildfires, and torching at least one town. Fruit was baking on the trees, and shellfish populations were collapsing as they simmered in hot tidal pools. Lenore had heard the media's call that it would be a hot girl summer, but she hadn't expected that label to be so literal.

The two of us exchanged notes and expressed admiration for the breathtaking scientific and logistical accomplishment of developing and administering COVID vaccines. We then expressed depression and frustration at the complete scientific and political failure to deal with climate change. Exhausting both topics, we returned to the serious business of feeding Mars.

Evan launched into a diatribe, "I think once we really start to plan out how we'll harness cyanobacteria and the microbiome on the Red Planet, it will change the whole way we think of farming on both planets. The microbiome makes us think small, compact, thrifty, and efficient. This is going to be completely necessary on Mars, of course, where every resource — hell, every square centimeter of inhabitable space — will be precious, hard-fought, and expensive. Living on Mars means living in a way where we waste nothing. But we also need to 'learn how to think small' here on Earth as well. For too long, we've associated farming with Big Things. Big fields, big business, big tractors, big herds and flocks, big yields. But as the wildfires, heat domes, and endless droughts show us, our species is pushing hard at the edges of what poor old planet Earth can handle."

Lenore nodded — we pretty much agree 100 percent with each other on these arguments. But by this point, Evan had built up a big head of steam, so Lenore leaned back, and let him continue ranting, figuring he'd get to the point eventually. When he did (a good five minutes later), she was ready to summarize:

"So, your point is that if we are going to survive as a species, we need to realize that the Earth too is precious, not to be taken for granted, and so we must become much thriftier in terms of how we go about the business of feeding humanity? Right? We need to learn to give more space over to natural systems?"

Evan nodded. "Yes! In other words, we will need to start thinking like a Martian: we need to think super-small. And this, Lenore, brings us into the wild world of nanotechnology."

For those of you not obsessively reading scientific bulletins, you will be forgiven for not realizing that a hidden revolution is happening in a field called nanotechnology. Sitting between the atomic, sub-atomic, and quantum scales, and the bulky material we humans call "the real world," is the nanoscale, which involves particles between 1 and 100 nanometers in size (by contrast, human hairs are generally between 80,000 and 100,000 nanometers wide).

Things really got going in the field in the 1980s when several breakthroughs, including a new generation of extremely powerful electron microscopes, allowed a trio of soon-to-be Nobel prize–winning chemists to discover a special kind of carbon molecule — called fullerenes — where individual atoms are linked together to create a nanoparticle shaped like a geodesic dome.[1] These particles are made up of many atoms but are still smaller than light waves. They are small enough to pass through almost any filter, and they never settle if they are suspended in a liquid.

Probably the most famous type of fullerene nanoparticle is the carbon nanotube, which (as the name suggests) is a tube of carbon where the walls of the tube are generally made up of just a single layer of carbon atoms, resulting in a cylinder only a few nanometers wide. Unbelievably strong, these tiny tubes are already used in the components of very fancy racing bikes and double-sided adhesive gecko tape. And for the history buffs out there, recent electron microscopic inspections reveal that the famous Damascus steel swords, which Muslim warriors used to great effect against Europe's crusading knights, contained carbon nanotubes. The nanoparticles found in the Damascus steel were presumably forged by medieval blacksmiths who used recipes that had been passed down for generations (and subsequently died out in the eighteenth century). Of

---

1   Curl, R.F., & Smalley, R.E. (1991). Fullerenes. *Scientific American*, 265(4), 54–63.

course, the blacksmiths would have had no idea what a nanotube was, only that when they used a special type of steel in combination with other key ingredients, their swords had a strength, flexibility, and sharpness that must have seemed almost supernatural to the European soldiers whose steel was brittle and dulled quickly.

Futurists believe that we could even braid long strands of carbon nanotubes to tether asteroids into near-Earth orbits and then use them as a gargantuan space elevator cable. Accompanied by a counterweight, this would create an effective conveyor belt that could simply lift material out of Earth's gravity using only a fraction of the energy needed by today's rockets to reach escape velocity.

Space elevators, Damascus swords, gecko tape, and fancy racing bikes aside, nanotechnology can make life better in the future in many other ways. In terms of food and medicine, one extremely promising approach is to use nanotechnology as a delivery mechanism that can take chemicals (fertilizers, medicines, or even extra DNA instructions) straight into targeted cells.

Today, most of this research is devoted to medicine, and one of the Holy Grails of nanoresearch is to put tiny amounts of cancer-fighting drugs into nanoparticle capsules and then slip those into tumorous cells to deposit the drug straight where it is needed. This allows tumors to be fought with greater precision, efficacy, and fewer side effects. Beyond medicine, there are a host of applications relevant in farming. As we shall see, getting these nanotools, along with harnessing the microbiome, will be necessary if we want to farm on the Red Planet. And if we're clever, we'll use these same tools to reduce the impact agriculture has on Earth's environment as well.

John Dutcher is a man who knows how to make a big deal out of a small thing. As one of Canada's leading researchers on nanotechnology, one of John's quirkier claims to fame is that he helped set up a start-up company out of his lab at the University of Guelph that produces nanotechnology-based skincare products. Essentially, the company uses nanoscale balls of a starch-like material called glycogen found in sweet corn to deliver moisturizer straight into dry, wrinkly skin. The phytoglycogen is absorbed by the body, leaving the moisturizer exactly where it is needed.

John and Evan are neighbors and normally see each other out walking. It was one of these chance neighborhood encounters that helped lead to this chapter. When they saw each other after a long gap, it wasn't long before they were engrossed in discussing all the possible ways that nanotechnology could make food production better and more efficient. Evan was looking for ideas, and John had, along with a large international network of top nanoscientists, just published a major review of the potential for nano to revolutionize farming.[2]

The most obvious way that nano will change agriculture is analogous to what's happening in medicine. Nanoparticles can be used as the delivery capsule for fertilizer and pesticides in much the same way as they can target tumor cells, fertilizing a plant with greater precision, thereby using less product and causing less waste.

Today, on average, crops' nutrient use efficiency (defined as the ability of crops to absorb and use nutrients) is only about 50 percent for nitrogen, 40 percent for potassium, and 20 percent for phosphorus. A huge amount of the fertilizer farmers apply is simply wasted. The stuff that is used by the plant helps boost yields, but the rest just flows off the fields where it either ends up in the atmosphere as a greenhouse gas or in the water table where it contributes ·

---

2   Hofmann, T., Lowry, G.V., Ghoshal, S., Tufenkji, N., Brambilla, D., Dutcher, J.R., ... & Wilkinson, K.J. (2020). Technology readiness and overcoming barriers to sustainably implement nanotechnology-enabled plant agriculture. *Nature Food*, 1(7), 416–425.

to problems like the cyanobacteria blooms described earlier. If nanofertilizers (or biofertilizers) could fix this problem, it would be a boon to farmers' bottom line on top of providing a massive environmental benefit.

In addition, nanoparticles can be designed to carry genetic instructions into plants as a way of enhancing pest or drought tolerance. And nanosensors are being used to allow plants to communicate with Wi-Fi-enabled smart farming systems to alert farmers if their crop is being attacked by pests or being stressed by droughts. These sensors would in fact be tiny particles that emit a minute signal if the plant is under stress. For instance, during a drought, the signals would be picked up and relayed via Wi-Fi or Bluetooth and alert the farmer to turn on the irrigation.

It all sounds fantastic, but these advances are being held back by regulators and consumers who worry that nanotechnology may have unintended negative consequences. For instance, some nanoparticles may accumulate in the environment and enter the food chain in a problematic rather than helpful way. So, it's important for scientists like John to do the research to understand the fate of these substances before they become too widely used.[3]

As the two walked, John said to Evan, "The consuming public and regulatory agencies alike are worried there could be problems if nanoparticles coded to do gene editing end up escaping into the wild. If that happens, would they hurt native species? We don't fully know yet, so one major aspect of the nano community's research effort is to ensure these tools are safe."

He's right. Any technology as exciting, and anything with as many wide-ranging applications as nano, has the potential to create problems. This means huge regulatory barriers, and years of careful

---

3   And this is one of the reasons that John works a lot with plant-based nano-particles — because they simply decompose naturally.

scientific experimentation, to ensure that the benefits of these tools don't outweigh the risks.

But maybe a Martian community might be able to help? If one worry is that by developing these tools on Earth, we'll make an even bigger mess of things here than we already have, maybe Mars is the ideal location to conduct this science? After all, there is no ecosystem to disrupt, and it is hard to imagine a more biosecure research facility than one separated from humanity (and Earth's rich biological heritage) by hundreds of thousands of miles of hard vacuum. Maybe Mars is the natural home for some of our most ambitious and sensational biotechnology experiments?

THE BIOFOUNDRIES OF MARS

"OK, so bring it all together for me," Lenore said. "Six or seven times a day, for the last month, you've texted me some interesting tidbit or factoid about nanotech and the microbiome. I get you're excited about Damascus steel swords and space elevator cables made of nanotech, but please lay it out, once and for all: How is this all going to feed our imaginary Martian community of BaseTown?"

"OK," Evan said after some heavy thought, "here is how it could work. On Earth, we'll need a combination of field experiments and big lab research where the goal is to uncover all the ways in which microorganisms and nanoparticles provide us with huge benefits. One output of this research will be a catalogue of things we've learned, like how nanoparticles can help us administer fertilizer precisely, or how the microbiome allows plants to remain productive in a drought, or whatever.

"But while this research will start at the field level, we'll need to understand the biochemistry, nanotechnology, and genomic elements of what makes these things tick. We will need to analyze crops, algae, bacteria, etcetera, and figure out which parts of a micro-biome's DNA are responsible for which sorts of traits. This will give us a baseline understanding — a sort of reference library or codex of biochemical, microscopic, and nanoparticle scale information about how to design that biological life-support system on Mars."

Lenore was familiar with other technologies in which key breakthroughs involved assembling vast libraries of plants, yeasts, and molecules. She was obsessively fond of libraries, and the idea of Mars holding giant repositories of knowledge appealed.

"Since much of this work needs to be done on Earth, however, we almost immediately hit the problem we started off with; it's your 'Jacob Two-Two Challenge' or the impossibly high costs and time it takes to get anything to Mars.

"So, as we learn how things work here, we'll need to digitize the genetic instructions for the microorganisms and the molecu-lar chemistry for nanoparticles. These will become digital blueprints that can be sent to Mars as data. There, we'll need biological manu-facturing facilities, called biofoundries, that will use gene-editing tools, microscopic 3D printers, and DNA synthesizers to build, or grow, the microbial inoculants or nanoparticles."

Evan's musings aren't as far-fetched as they sound. Scientists have already built yeast organisms out of basic elements, and lots of scientists are taking the raw ingredients for life and building new things out of them. In fact, a lot of these technologies went into developing COVID vaccines.

The basic building blocks of life — DNA, proteins, amino acids — are made up of common elements, all of which can also be found on Mars. Amino acids, for instance, are made up of hydro-gen, nitrogen, carbon, and oxygen. The chemical structure of DNA is four nucleotides (the basic structural unit of nucleic acids) that are made of nothing more exotic than phosphorus, oxygen, nitrogen, hydrogen, and carbon. Carbon nanotubes are just carbon, of which there is plenty on Mars. The elements used in John Dutcher's nanoparticles are similar — they are mostly just carbon, oxygen, and hydrogen. While these elements aren't necessarily abundant all over the universe (which is mostly made up of hydrogen and helium), they are abundant on both Earth and Mars.

So, we don't need to get lucky and find lots of rare elements or fancy chemical materials to sustain life on Mars. Rather, we need to learn more about Earth. The bewildering diversity of life we witness is made up of some relatively commonplace atoms. Here the scientist's job can scan the genetics of plant life and then show how these basic building blocks can be assembled into new things — new nanoparticles, new forms of protein.

We can use the advances in the microbiome and nanotechnology to develop self-sustaining biological life-support systems that use Martian ingredients to produce a viable organic system, but we can also use that same life-support system to test new tools that will help us manage Earth's systems better than we currently are doing. In time, probably not a lot of time, Martians would be returning the favor, sending back knowledge of their own discoveries.

Evan's thinking made sense to Lenore. "I like it," she said. "I mean, it's 90 percent science fiction, but with serious talk all over the place about sending humans to Mars, we need to be thinking out of the box. And I like the way you've linked BaseTown and Earth, where this planet does the basic blueprinting and Mars manufactures things based on digital plans. In fact, I think I love it. But for all the work you've done, we still haven't figured out how a Martian Jacob Two-Two is going to get his two pounds of firm red tomatoes. Luckily, all this time you've been texting me about nano and micro, I've been reading up on how to grow things people will want to eat. Let's wrap up this small-scale work and go talk about tomatoes. I'm hungry."

PART II:

RED EDEN

CHAPTER 4:
Biophilia

TITAN ARUM

While Evan delved into the world of the very small and even smaller, Lenore was looking at things of a botanical nature. It was a rainy spring day in Vancouver, and Lenore was feeling restless. She decided to stretch her legs and scratch a bit of the wanderlust itch with a visit to her friend, the Java finch. These small, posh birds, dressed in grayscale finery, would not normally live in Vancouver; they are native to Java, Bali, and Bawean in Indonesia, where they feed on grain and seeds. They particularly enjoy rice fields, lending them their Latin name, *Lonchura oryzivora*: the rice eater.

Vancouver is not Bali, so why is such a hot weather bird flitting about? Lenore's feathered friends live in the Bloedel Conservatory, a bubble of warmth, aluminum, and plexiglass that holds back the cold and rain. Conservatories, and their workhorse cousins the greenhouses, are one of Lenore's passions. The Bloedel Conservatory is a modest but elegant example, perched upon a rise in South Vancouver's Mount Pleasant neighborhood, set atop the tallest hill inside city limits, the ironically named Little Mountain.

The conservatory was a Canadian centennial project,[1] championed by park superintendent Stuart Lefeaux, who saw an opportunity to transform the lid of a five-acre water reservoir into something wonderful for the community. The Bloedel Foundation contributed $1.25 million, the largest donation to the city at the time.

The Bloedel money paid for a delightfully retro-yet-space-age triodetic dome, forty-three meters in diameter and twenty-one meters high. The dome has 1,490 plexiglass bubbles, in thirty-two different shapes, and 2,324 pieces of aluminum tubing. The style (which was all the rage in the 1960s thanks to Buckminster Fuller) was designed and patented in Ottawa, fashioned to be rigid and strong and resist extreme weather. No interior columns were needed, and even sixty years later, the dome still feels futuristic, glowing warmly, like a kind of alien overlord, benevolently ruling the city below.

Lenore likes the architecture, but the real treat lies inside. The dome contains three tiny biomes: a tropical rainforest, a subtropical rainforest, and a pocket of desert. Two hundred bird species flit about these habitats, and an array of rare fish dot the ponds. There are five hundred plant species, including bougainvillea, citrus, coffee, eucalyptus, and hibiscus. On a dark, rainy day, armed with a good book, the dome is cheerful and relaxing, and Lenore enjoys the deep smell of plants and water while basking in the bright artificial sunlight that contrasts Vancouver's winter gloom.

There is even a titan arum named Uncle Fester. More commonly known as the corpse flower, our largest flowering plant is famous for putting up an enormous three-meter-tall nocturnal bloom every few years that, once open, fills the air with the smell of rotting flesh. On

---

1   For the non-Canadians, centennial projects were projects launched in 1967, on the occasion of Canada's centennial.

non-flowering years, corpse plants simply grow a single four-meter-high leaf, slowly gathering energy in their seventy-five-kilogram corm for the main event. In July of 2018, Lenore and her friends made a night of it and lined up outside the dome with other curious Vancouverites to see Uncle Fester in action. It was, indeed, a big flower with a big smell.

This story made Evan laugh.

"I don't get it. A big part of your professional life is to go to greenhouses covering dozens of hectares, full of cucumbers, tomatoes, lettuce, and run by a whole array of space-age technology. And yet you are willing to give up a night's sleep to see a stinky plant in a 1960s building? Isn't the conservatory sort of a toy in comparison?"

"It's true, but I think even on Mars we will need not only sustenance but beauty too. The need for food will be so intense that in many ways the whole of any Martian community will need to be a greenhouse. And while this might take on the feel of a factory, it should really feel like a garden.

"The point is that the spaces we build on Mars will be so scarce, and the outside so harsh, that many of the indoor spaces are going to have to be both functional and pleasant. We may be able to bury some bioreactors to produce algae, but we really won't have the luxury of hiding dirty industrial zones away and out of sight as we do here on Earth. So, I think the whole place is going to have to be built to the highest aesthetic and functional standards. And most people like gardens because they are both pleasing and practical, like the kitchen garden at Hampton Court."

Lenore's right. The practical and the pleasing have always shared space in well-designed gardens. And creating small, enclosed areas that will serve for both recreation and cultivation will be vital on Mars. But while we are looking to the future, it's important to realize that cultivating plants indoors stretches back to the beginning of civilization, a coming together of agriculture and architecture to create bounty and beauty.[2]

THE MOODY EMPEROR'S CUCUMBER

There once was a moody emperor, named Tiberius Caesar Augustus, whom historians have accused of all manner of excesses and crimes. Perhaps in an effort to soothe his turbulent emotions, Tiberius's physicians recommended the emperor eat a daily dose of fresh cucumber, and so the gardeners at his Villa Jovis on the Isle of Capri[3] developed an early prototype for greenhouses, called specularia. These were fashioned as small carts, filled with compost and covered with thinly cut sheets of selenite, a translucent rock that allowed sunlight to enter but prevented heat from escaping. The carts were then moved (presumably by slaves) inside and out, depending on the time of day and weather, thus supplying old Tiberius with his daily fix.

Whether these were the first greenhouses in history used to grow cucumbers, we will probably never know. But the idea, even then, was far from new. People had been stretching growing seasons

---

2   Gardens, and more importantly the enclosed gardens that lead to the controlled environments needed for a Mars mission, have their roots in cultures and religions from around the world. For the purpose of this narrative, we are going to follow the Roman line of development, but please be aware that creating habitats for plants is a global pastime. However, from prehistory to the first attempts to sprout seeds in space, the goal has been the same. As a species we have a desire to grow plants locally even if the environment is not suitable for their growth. To do so, we have to get a little clever.

3   Gardeners take note: the villa also featured a spot known as Tiberius's Leap, a shear drop to certain death. The emperor was known to toss over servants who weren't achieving what he desired or houseguests who overstayed the emperor's welcome.

for centuries by training tender plants up south-facing stone walls that held the heat of the day.

Whether the cucumbers soothed the emperor's nerves is also not known. In fact, we don't even know if it was cucumbers they planted. The Roman historian Lucius Junius Columella's description of what was planted in the specularia suggests that the plants may have been snake melon. Snake melon, which resembles the cucumber, is much more of a medicinal vegetable being related to the musk melon from the Middle East and sometimes called a gutah. The story may be more about desperate healers tending to a gravely ill ruler by cultivating rare imported medicinal plants than anything literally about cucumbers.[4]

What is known, however, is that despite specularia being written about by both historians Columella and Pliny the Elder, this technology didn't really catch on. It was an idea before its time, made possible by the emperor's vast wealth, eccentricities, and pool of slave labor. So, for a while greenhouses drop out of the historic record.

It wasn't until the 1450s that we have solid evidence of a heated greenhouse, the logical next step. This advance took place in Korea and is described in the *Sandu Yorok*, a book written by a royal physician during the Joseon Dynasty. The text describes under-floor heating creating spaces to grow out-of-season vegetables and force flowers. The walls of these were thick, and the windows were made of oiled paper. These structures could, apparently, even produce oranges in winter. But despite Korea's advances, the whole notion of indoor growing still didn't take off. Neither Korean nor Roman models had what it took to make a real impact on the food system. It wasn't until the Victorian era,

---

4   Desperate, of course, as we have to imagine he was still capable of having people tossed off of the cliff.

with showstoppers like the Crystal Palace and the proliferation of Victorian conservatories, that we start to see the idea of growing food indoors really take hold and spread.

The Romans and Koreans had the idea for a greenhouse but lacked a critical ingredient that kept the technology from taking root, so to speak. In both cases, they needed a substance that could hold in heat while allowing light to pass. Plants need light, and the options available to the Roman and Korean gardeners — oiled fabric and thinly split translucent rock — just didn't cut it to make indoor growing practical.

The point is that, although the idea of the greenhouse was there, certain technologies still had to develop before growing plants indoors could advance. To do greenhouses properly, gardeners needed a cluster of technologies such as the right plants, lightweight building materials, and heating systems, each of which had to be practical and available before greenhouses became viable. All these different types of technology must have analogues on Mars. So, to explore how we might end up with beautiful and sophisticated greenhouses on the Red Planet, let's begin with the basics and take a technological sidebar into the history of one of our most important materials: glass.

LET THERE BE LIGHT

It's hard to imagine life on Earth without glass. But even though it has become a ubiquitous part of our daily lives, it is a surprisingly recent invention. This might seem especially odd given that soda-lime silicate glass, the common glass that fills our world, is literally made from silica, the primary component of everyday sand. Sand melts at the very high temperature of 1,700 degrees Celsius, and at that point, it doesn't cool back into the grainy yellow beach sand we are used to. Instead, it forms what is called an amorphous solid, a cross between a liquid and a solid. This new material, glass, is also transparent.

Heating sand to this melting point is a challenge even today, and so we mix sand with soda ash (sodium carbonate, also called flux) and limestone (calcium carbonate).[5] The soda acts as a catalyst and radically decreases sand's melting point. The limestone is a fixative; without it, the resulting glass would dissolve in water. This sort of glass has been made for a very long time, in the form of drops and beads that have been found in archaeological sites dating back to at least 2000 BC.

Pliny the Elder, one of the Roman historians who wrote about Tiberius's specularia, tells a cute (though perhaps apocryphal), story about the discovery of glass. In his telling, a ship filled with niter moored for the evening near Lebanon. The merchants aboard built a fire pit out of niter lumps to hold their cooking pots and were soon shocked to find streams of an unknown transparent liquid pouring from the fire. They had discovered that a flux can catalyze the melting of sand. The Romans also used niter and vegetable ash to cure the glass and make it waterproof. It was this basic recipe of sand plus ash filled with sodium and potassium that would form the bedrock of glassmaking until the Victorian era.

This approach only created lumps of glass and a lump of glass is of little use save for beads and ornaments. In the first century AD in the Middle East, artisans began using a long iron pipe to blow bubbles of glass to create bottles and containers, and we see many such

---

5   While it seems that carbonates are relatively rare on Mars, it may be possible to synthesize these compounds with readily available ingredients rather than relying on too many supply runs from Earth.

objects from Roman times. Flat panes of glass, however, were more challenging. Romans made crude window glass by pouring molten glass onto stone slabs and then frantically spreading the cooling lump. Such window glass was crude, small, and translucent, but it was still prized. Examples were found in Pompeii and Herculaneum.

In the second century AD, a much better way of making flat panes of glass emerged. Master glassblowers created very long, wide bottles, cut off the top and bottom, and then opened the remaining cylinder of glass into a flat pane. This method was hit and miss but was used for centuries. Glass panes remained rare into the fourteenth century, the skill of crafting them passed down in glassmaking families.

During the medieval period, another way of making glass emerged; lumps of molten glass were spun at speed until they flattened into discs as much as three or four feet across. One- to two-foot panes could then be cut out of these discs of crown glass. Crown glass was usually thicker on one side, but it was clearer than previous products, and with its advent, greenhouses and conservatories finally began to appear, though they were still limited by their cost.[6]

Plate glass finally arrived in the Victorian era, formed by pouring molten glass onto long iron tables and machine rolling it. This new glass was then ground flat, smoothed, and polished. It was a costly process that still didn't lead to the clearest product, but plate glass panes could be very large and were often used in early railway stations, thus allowing glass to become a serious architectural element for the first time. This new glass was still thick, heavy, and eyed with suspicion. Hence, even the great halls of the 1851 Crystal

---

6   Contrary to popular myth old glass doesn't slowly flow over time. Old glass is thicker on the bottom because that is how the naturally uneven crown glass was mounted. Once cooled, glass is as stable as a solid.

Palace in London were clad in crown glass that was manufactured in ways little changed from medieval times.

The next big breakthrough — and the breakthrough that has given us the look and feel of the modern world — came in 1952, when Alastair Pilkington, of the Pilkington glass company, devised a cunning new approach to glassmaking. He pioneered the creation of float glass, where a continuous ribbon of molten glass emerged from the fires and was floated on a perfectly flat trough of molten metal (usually tin). By 1958, this method allowed for the creation of near-perfect glass — a long way from a puddle of melted sand emerging from a fire pit lined with niter.

This ancient technology matters for Martian communities as they need to be self-sufficient.[7] They will need to use what they have, and that suggests they will be making glass out of Martian sand.

Plastics and plexiglass are based on petrochemicals, but on Mars, petrochemicals will be scarce. Some petrochemicals will presumably be made using advanced fermentation in bioreactors, but for the amount of material needed for Martian greenhouses, we will need to rely on what the Red Planet itself can provide: Martian sand to make Martian glass. It's cheap, it blocks some radiation, and it is easy to make.

Lenore was taken by a strange idea. Domed craters on the Red Planet might look like the Bloedel Conservatory, only made of Martian glass. She wondered whether there would be Java finches.

More technically, the folks in BaseTown are going to have to tint the glass to filter out harmful radiation and use thick enough panes to prevent the stuff from rupturing if there is a meteorite strike. As well, the habitats are going to have to hold enough pressure for humans to tolerate, and they are going to need to triple

---

7   The space wonks call this *in situ resource utilization*, which is just a fancy way of saying Mars needs to produce what it needs.

glaze everything to keep out the cold. Luckily, some of NASA's rovers have shown that 1 percent of Mars's atmosphere is made up of the element argon, so this means that future Martians may be able to extract this argon and inject it between the panes of glass just like in a modern Earthling window, to make it super well-insulated.

Lenore picked up the thread. "So, there are tremendous challenges to designing the right kind of greenhouse, and the look of the place could be a sort of cool retro steampunk, but the foundations of the community will be pure science."

GOLD, DIAMONDS, AND FRUIT

While conversations about possible architectural aesthetic styles for Mars may seem trivial (or perhaps even odd), it's not. A key theme we will keep returning to is that the built form — the design of our tools, buildings, and communities — must consider more than the function it is to provide. We must also strive for beauty. We should strive to make things appealing to the eye, the mind, and the soul as well as being designed to be functional.

Designing good-looking things won't be a luxury but a necessity on Mars. Life will be incredibly tough for the early inhabitants, so ensuring that they enjoy a beautiful living arrangement will help maintain the mental health of these pioneers (similarly, another theme of this book is that the food must taste good, as a sure-fire recipe for disaster is to assume that the food the Martian residents will eat will be optimized for nutrition alone). Building beautiful things is something we should do more here on Earth, too, where we often put function over form, transforming major swaths of our planet into pure ugliness. Here again, we can draw lessons from the history of the greenhouse and the conservatory;

the popularity of these structures seems tightly linked to the human desire for bling.

As the Renaissance unfolded, it seems that everyone who was anyone on that soggy island known as England tried to produce, at great expense, the odd orange or lemon in a stately and elegantly appointed conservatory. Want to see the pomegranate that sent Persephone to Hades leaving Demeter to mourn? The local lord and lady of the manor can set you up if you are posh enough to warrant an invite to the music night in their new conservatory. Can't afford the equivalent of $8,000 for a pineapple? Consider a rental! Welcome to the wild world of the conservatories for the rich and famous of the Early Modern World.

As crown glass became available in quantity, the conservatory became a true outdoor room, producing rare fruits and blossoms. Much as the Romans did with their specularia, these early iterations of the modern greenhouse were often heated with decomposing dung, that warmed plant beds as it broke down. (This also explains why the conservatory was often separated from the main house by a vestibule. The smell of rotting manure must have been most unappealing in the ballrooms and dining halls.)

The Europeans also reproduced the Roman use of chimney heat to warm plants, running flues through the north walls of the glassed-in enclosures. The craze began with citrus. The orange made its way from Asia, where it grew wild in southern China, over to India, and then up into the Roman Empire. But it remained uncommon in Europe until the Portuguese began importing the fruit in bulk during the Middle Ages. In Northern Italy, citrus trees were grown in containers that could be wheeled indoors (or into caves) for winter. The Villa Palmieri, just outside of Florence, Italy, has grown lemons and oranges continually since at least the 1450s. Its current Lemon House was built in 1700 and used as a model by French aristocrats who fell in love with citrus as they toured the wonders of the classical world. Some of the very trees at the Villa Reale di Castello date to the time of the Medici, showing the true sustainability of enclosed growing.[8]

Lenore once spent about twenty-five days in August walking the London Greenbelt, covering a landscape that included Southend, London, and Canterbury. It rained every single day. Evan, a dual UK-Canadian citizen, spent many years living in the UK, struggling to waterproof his home and family. Both of us can attest that the UK is not citrus country. But rising to the challenge, the English aristocracy of a bygone era worked desperately to outwit the weather. At Beddington House in Surrey, Francis Carew built an enormous citrus collection in 1539, and each winter he constructed a heated structure of planks and canvas around the trees. In a particularly showy demonstration of his horticultural skills, he held back the fruit of a black cherry tree by packing it in damp straw so that the fruit would be ripe in August of 1599, timed perfectly for the visit by Queen Elizabeth I. Some of his techniques still ring down through the ages and will likely be used on Mars. His orchard survived for more than a century, producing pomegranates, lemons, and up to ten thousand oranges a year.

Across the channel, in France, an even more impressive display of controlled-environment growing was developed at what had once been a modest royal hunting lodge. The Royal Orangerie at Versailles was designed in 1685 and is still impressive today. The Sun King kept his trees in silver tubs to be wheeled in and out of doors and provide visual interest and pleasant perfume. The Orangerie is double glazed with crown glass, an astounding feat allowing the Orangerie to function with little auxiliary heating.

---

8   Assuming, of course, one has the funds for both fuel and labor!

Back in England, the seventeenth and eighteenth centuries saw impressive expansion of controlled growing environments. The horticulturalist John Evelyn first used the words conservatory and greenhouse in 1664 and developed new structures and a novel heating system using pipes under the floor. The University of Oxford developed an impressive Physic Garden of medicinal plants, and new architectural techniques provided higher temperatures in conservatories suitable for both sugar cane and the coveted pineapple. In another part of the UK, the impressive Dunmore Pineapple still survives today. This eighteenth-century stone building is — as the name suggests — shaped like a pineapple, and it produced ripe pineapples in Scotland.

Then there is the great vine at Hampton Court, planted by the master century gardener Lancelot "Capability" Brown, that still bears a heavy crop of grapes each year. Perhaps, one day, some future Martian might sit beneath a similar vine, sipping her own exotic vintage.

If the English excelled at the showy, the Dutch mastered the practical. The botanist Charles Bonaparte built what is often described as the first modern greenhouse in Leiden during the nineteenth century to produce medicinal plants. Dutch universities also began to produce their own collections of plants grown under controlled environments, creating a research enterprise still active today.

While these early conservatories nurtured exotic plant life, the development of cheap and strong cast and wrought iron in the Victorian age took their architecture to new heights. Builders could raise ever more elaborately beautiful structures, including those that can still be seen at Kew Gardens, the famous botanical garden near London. For the first time roofs could be entirely glass, often made of overlapping panes of crown glass in a style known as scalloping that could be fashioned into curves and domes. At Kew, the Palm House was heated with hot water piped under the floor, creating such a pleasant environment that Victorian men and women lounged there to read the news and take chocolate. In one of the finest examples of the art, Joseph Paxton assembled the Crystal Palace in record time for the Great Exhibition of 1851. Three storeys tall, it dominated its corner of Hyde Park. Thirty-eight hundred tons of cast iron, seven hundred tons of wrought iron, and 900,000 square feet of glass — one-third of Britain's annual glass production at the time — were used in its construction. Six million people visited during the fair.

Domes will play a role in Martian design. They enclose the largest possible volume for the surface area of material. Domes will enclose some of the agricultural capacity of a Martian community.[9] And if we think of enclosing full-sized trees on Mars, to create a sense of open space, the dome is the way to do so. Such domes might well follow the design of thinker R. Buckminster Fuller, the man who nearly single-handedly associated domes with futurism and space exploration. Born in 1895, Fuller had a decidedly unorthodox life. He was expelled from Harvard twice, losing interest in his studies in favor of partying with a vaudeville troupe. He went on to work as a mechanic at a meat-packing facility, and served in World War I as a naval radio operator. It was during the war he became fascinated with the need for fast, strong, modular construction. How could we cheaply and easily create reproducible structures

---

9   When Lenore thinks of domes, her mind often turns to the Roman pantheon, a temple to the gods turned Catholic church that has been continuously occupied since 126 AD. It is dominated by a massive concrete dome topped with an oculus, a metaphorical eye of god that floods the space with natural light. And here is the strange thing about the pantheon; it is still, after millennia, the largest unreinforced concrete dome in the world. The pantheon feels (and is) a vast space.

that would work in any climate? His first attempt at creating a business in the modular housing arena failed, and the death of his daughter from polio threw him into depression and alcoholism.

Many people's stories would end at this low point, but Fuller found inspiration and recovery as an instructor at a North Carolina college. He rediscovered his love of theater, which gave him the confidence to speak in public and to reengage with the problem that defeated him earlier: modular construction. This time, Fuller developed a dome constructed out of hexagons and pentagons.[10] He patented his work and perfected his plans in 1949. His final products were extremely strong and, key to space exploration, could withstand forces both from outside and inside the dome. The U.S. Marine Corps bought thousands of his domes, making Fuller famous.

As a result of Fuller's work, it's likely that giant glass domes may feature prominently on the Martian skyline. They would have to be tough, able to withstand both the meteorite strikes that are common on Mars due to the thin atmosphere and the force of being pressurized, tinted to let in sunlight but filter out harmful radiation, and extremely well-insulated. But it's possible to imagine engineers overcoming these challenges with materials that can all be found on Mars and with technologies that are mostly already available.

Perhaps, small domes will be used at the end of tunnels as habitation caves on cliff faces and provide warm, naturally lit sitting areas for community meals. Maybe other domes will arch gracefully over craters to create spaces to grow grain crops and provide a bit of green space where Martian community members might gather to listen to concerts much as the Victorians did.

Maybe one dome might even be built over a crater close to the north pole filled with frozen water. Solar collectors and mirrors

---

10 Each geodesic sphere must have twelve pentagons.

could orbit above this spot, programmed to focus the sun's light onto this particular dome, warming the enclosed space, melting the ice, and turning it into Mars's first natural lake (well, the first one in a long time). Some master gardener might follow in Capability Brown's footsteps, creating a citrus grove by a beach, only a few hours' train ride from the main Martian community.

If we let our imaginations run wild, we can start to piece together a setting that might be a little like the Eden Project in Cornwall. Nestled in an abandoned clay pit on the English coast between Penzance and Plymouth, it looks very much like an off-world community, or a collection of oversize soap bubbles. The clay pit had been in use for nearly two centuries and was so degraded it became the set of the ruined planet Magrathea in the televised version of *The Hitchhiker's Guide to the Galaxy*.

Opened as a botanical garden in 2001, the site grew as plants were slowly sourced from around the world to showcase the fact that with enough ingenuity life could be coaxed to grow in even some very inhospitable places. These plants were placed in two specially designed biomes. The tropical biome, which covers 1.56 hectares and is fifty-five meters high, is filled with plants such as bananas and coffee. The Mediterranean biome is 0.65 hectares and thirty-five meters high. It holds the sort of plants the Victorians enjoyed, such as olives and grapes. The domes themselves are geodesic structures of tubular steel and thermoplastic panels. The Eden Project, though, wasn't just designed to be pretty. It was designed to show how plants are interdependent, how they need each other to survive. The same will be true on Mars.

"Hang on," Evan said as he absently doodled geodesic domes on the whiteboard of our shared Zoom window (not an easy task, nor particularly effective). "I understand why you think there will be conservatory-style domed spaces on Mars, and how they will produce plants, but here on Earth the largest, most productive greenhouses don't look like that. They are literal plant factories — and are total function-over-form buildings designed to optimize production with as little human intervention as possible. In fact, to keep crops free from diseases, pests, or other problems, humans are kept out except for maintenance, and generally, these things are run by robots."

Lenore did know this; one of the best parts of her job was her access to some of the most productive greenhouses on Earth.

"That's absolutely true," she replied. "The most stunning spaces on Mars will have plants, and those plants will produce food, but I think you're right — the bulk of the plant food will need to be grown elsewhere. After all, the microorganisms that will form the basis of the Martian food system will likely be grown in big vats rather than greenhouses. And most plants can tolerate lower pressure and less oxygen than humans but grow best with more carbon dioxide than humans like.

"So, while human habitats will all have plants, the reverse may not be true, and places designed for plants will not be kept at conditions that would be good for humans. For example, I think fruits and vegetables might be farmed in tunnels or caves, kept warm, enriched with carbon dioxide, but low in pressure and oxygen (in fact, the Martians would likely try to harvest the oxygen the plants produce). These caves would probably have almost no human presence at all, and when anyone does go in, they'd need an oxygen tank and full biohazard suit to keep the place sterile. In other words, the productive parts of the Martian communities are going to be autonomous plant factories, while the places that are used for human habitation will also be full of plants."

Evan nodded. "Yeah, OK. So cool neo-steampunk domed conservatories for living, and these also produce a bit of food. But the bulk of food production comes from super intensive biofoundries and plant factories off to the side. This all makes sense to me, so let's talk about these plant factories."

SUMMER IN A BOX

Earlier, Evan and Lenore had both taken an online course offered by the Japanese Plant Factory Association to better understand the technologies and science behind the world's most advanced indoor farms. Plant factories fall under the category of vertical agriculture, or more technically "intensive controlled-environment growing." This is a direct offshoot of the modern production greenhouse, which really came into its own in the rainy gloom of the Netherlands. Greenhouses cover well over ten thousand hectares in that country, mainly in the Westland region, which has more concentrated greenhouse production than anywhere else in the world. Some Dutch greenhouses are even built on giant floating barges.

The Japanese saw an opportunity to build upon the success in the Netherlands; after all, Japan also has plentiful water, electricity, long dark winters, and a technologically trained workforce. The basic principles of the Japanese plant factory are that space and resources are used to their utmost, and every variable is controlled. In a general sense, plant factories are automated, artificially lit greenhouses that produce fruit and vegetables year-round by controlling environmental conditions such as light, atmospheric composition, temperature, humidity, air pressure, and nutrients.

The controls of the plant factory are highly automated, with computer programs managing virtually all aspects of the growing environment. The resulting agricultural products are safe, organic, uniform, and — as they are produced close to markets — extremely fresh. As we did virtual tours of plant factory operations, the two of us noticed that the plants seemed to glow with life and health. This is not your grandfather's lettuce. In the plant factory, poor harvests and bad years are rare, and late frosts and hailstorms inconceivable. Even pest outbreaks, or attacks by fungal pathogens, seldom happen as these factories are biosecure, cut off from the outside world. Hence, there is virtually no need for pesticides.

The origins of these plant factories stretch back to the post–Second World War period, when Japanese engineers riffed on and expanded a new technology developed at Caltech called a phytotron (basically another word for a research greenhouse). The first growth chambers in the early phytotrons had nothing more than some crude controls to change lighting, temperature, humidity, carbon dioxide, wind, and mist. While interesting for researchers working on plant physiology, the technology didn't catch on commercially in land-rich Southern California. After all, why build a phytotron if fields are available cheaply? But in Japan, where the amount of arable land per person is much lower than in the U.S., universities began tweaking the approach, making it better.[11] Progress was slow, and getting efficient plant factories on the market was held up by one limiting factor or another. But by the first decade of the twenty-first century, technologies matured and the tools to control temperature, humidity, airflow, and hydroponic systems made sufficient progress that crops like tomatoes, peppers, and cucumbers grown under glass became ubiquitous. Lighting these controlled-environment settings was always tough. Fluorescent bulbs helped improve production, and Japanese

---

11   At roughly the same time, Japanese automotive engineers were doing the same thing with car manufacturing, essentially redefining the way the world made cars and causing huge upset to North America's car industry. While it's still too early to tell, both of us think that history may be on the verge of repeating itself, and it's possible that these Japanese plant factories are about to redefine how fruits and vegetables are produced. North America's established produce industry should pay attention to this historic precedent.

farms began stacking layers of plants on top of each other, making early plant factories more efficient by growing up rather than out, but for decades it seemed nothing could improve on sunlight.

Nick Holonyak Jr., an engineer at General Electric in Syracuse, NY, didn't set out to change the world in 1962 when, at the age of thirty-three, he created the first practical light-emitting diode. He wasn't even trying to create light, but rather a better laser. Semiconductor lasers were still theoretical, but Holonyak knew that some semiconductors lit up when an electrical current was applied. As he experimented with various semiconducting proto-types, he found a combination that gave off a bright red glow. Dubbed *the magic one*, this simple gallium arsenide phosphide alloy created a tiny bulb that General Electric quickly turned into a practical lamp that they sold for a jaw-dropping $260. A green bulb followed, and, ten years later, one of Holonyak's students mastered a yellow one. Even at high prices, these bulbs lasted longer, were smaller and cooler than existing lights. And, as each year passed, they became better and exponentially cheaper.

Finally came blue lamps, a breakthrough pioneered in 1994 by Shuji Nakamura that allowed engineers to create a white LED that could replace the world's incandescent bulbs. This was a key-stone technology if there ever was one. Developing blue LED built on three decades of innovation with light-emitting diodes, but until this piece of the puzzle dropped into place, the rest of these innovations didn't matter all that much. Blue LED was such an important technology that Nakamura and his team were awarded the Nobel Prize in Physics, and since that moment, folks working in all manner of scientific and economic disciplines real-ize there was a time "before blue LED lights and after." It was a moment of singularity, and the following decades have been about improving quality, lowering cost, and widening applications to replace almost all other light sources.

Mars will run on LEDs. But even today on Earth, the revolution in lighting is changing the world we live in. Mass application of LED technology is saving money and transforming how lighting works, limiting both waste heat and e-waste. They also last much longer, a critical feature when we consider the need to conserve resources on both Mars and Earth.

LEDs are the lighting of choice in plant factories. As they are cooler, they can be closer to the plants, and as they can be targeted to specific spectra, we can tailor the light wavelengths each plant receives, radically changing the final product. Leaf lettuce, for example, grows well in the early days after germination when exposed to a bit more of the blue wavelengths. But if a farmer adds more red wavelengths in the last few days before harvesting, the plant is flooded with antioxidants that purple the leaves and that consumers love. By studying how plants grow under different wavelengths, and matching these experiments with plant breeding programs, plant factory operators are creating the specialized crops optimized for these conditions as well as "light recipes" that can help change the size, flavor, and color of each crop.

These innovations spurred the Japanese government to launch their Economic Growth Strategy for Widespread Plant Factory Use in 2008. At present, hundreds of plant factories are operating in Japan, providing high-quality products while continuously improving their equipment and efficiency. This revolution is just beginning to reach North America.

Japanese plant factories were closely associated with a series of driving forces: food quality, environmental outcomes, price, lack of land and labor, water conservation, and climate goals. These Japanese plant factories followed a cycle that will also be critical on Mars. First, germinating seeds and growing seedling plants in one location, and then moving the plants to a different location within the plant factory where conditions are right for finishing.

To explore the potential for this in a bit more detail, Evan turned to the real experts: the scientists who are right now designing both the experiments and the life-support systems that will literally sustain the next generation of lunar and Martian missions. Luckily, Evan works at the University of Guelph, which refers to itself as Canada's Food University, and this means that there's not just one but two world-renowned experts on campus working with the Canadian Space Agency, the European Space Agency, and NASA to tackle this exact problem. Based at the Controlled Environment Research Facility (itself a space-age lab full of growth chambers, LED lights, and banks upon banks of plants) are professors Mike Dixon and Thomas Graham. Mike describes himself as a space explorer and is determined to be the first person to grow barley on the moon; Thomas cochairs the selection jury for the Canadian Space Agency's deep space food challenge, which is looking at novel ways of sustaining long-term space missions.

It wasn't long into their chat before Evan and Thomas were swapping geeky science stories about amazing discoveries about vertical farming. They also feel to discussing plants can survive at relatively low pressures (which will be beneficial when it comes to growing plants off-planet), but that this causes problems for pollinators such as bees that often can't fly in these same low atmospheric pressure conditions. So, while the plants may grow okay in a lunar habitat, pollinating them becomes a major scientific challenge.

And then Thomas told Evan about the most technologically sophisticated vertical farm known to humanity. At the time of writing, this is still hypothetical and is part of an international effort to become the first farmers on the moon, but the first small steps towards this giant leap are in the works with a consortium of researchers and technology providers working to grow plants on the moon in 2025.

The challenge is easy to describe but incredibly complicated to solve. Can you germinate seeds on the moon and keep the resulting plants alive for one lunar day and one lunar night? One part of the challenge is the fact that on the moon a day and night can last as much as two Earth weeks. Another aspect of the challenge is the fact that during the day, the temperature on the moon can soar above one hundred degrees Celsius while at night it may plummet to less than minus-150 degrees Celsius. Then there's all the solar radiation to contend with.[12]

The plan Thomas and colleagues have come up with is really nothing more than a small well-insulated box designed to keep the temperature and humidity relatively stable. The whole thing will weigh only ten kilograms and be designed to hydroponically grow *Arabidopsis*,[13] barley, and another yet to be determined crop. Some of the plants will be genetically tagged so that when the lunar environment stresses the plant, and the plant responds by turning on certain genes, the plant will also produce a green fluorescent protein (originally from a jellyfish, but that's another story) that the team can image to see where and how the plant is responding to the lunar environment. Tiny multispectral cameras, built into the growth chamber, will take reams of photographs as the plants germinate and grow, sending that information back to Earth, thereby letting scientists like Thomas and Mike explore how and why plants on the moon become stressed and how they respond. If all goes to plan, this will happen in 2025 on the south pole of the moon and will be

12 Note: the lunar south pole, which is where this experiment is planned, isn't quite as extreme as this.

13 Also known as rockcress, a brassica plant in the same family as cabbage and mustard, this type of plant is often used in plant science research given that it is relatively easy and cheap to grow yet acts like most other flowering plants. See: https://www.nsf.gov/bio/pubs/reports/arabid/chap1.htm

an incredibly important step to developing the life-support systems humanity will need as it looks to the stars. As Thomas describes, "We need to figure out how to keep the plants alive in space because the plants are what will keep the people alive."

GLASS CASTLES

Lenore sketched a leaf on the virtual whiteboard. Then, she added a dome around it and an arrow coming in from a box labeled *bioreactor* and an outward arrow to another dome called *human habitat* + O2.

"This is basically a rough schematic of the BaseTown architecture we've devised. The point is that, ultimately, life on Mars will rest on these biological processes. On the plant side, the leaf is the primary site of photosynthesis where energy and carbon dioxide are turned into oxygen and food, usually carbohydrates. It doesn't matter if it's a pine needle, or an asparagus scale, or a cactus spine, or the cyanobacteria, every cell capable of doing photosynthesis is a frontline autotroph on Mars and Earth.

"So, we have chlorophyll absorbing light, producing food and oxygen, another cornerstone of the Martian ecology and economy. Plus, the transpiration of plants on Mars will keep the humidity up. Then we have the light. As we've talked about, Mars is farther away from the sun, so they'll need a combination of mirrors, lenses, and fiberoptics to concentrate sunlight, and this will augment the LEDs that will be all over the place.

"Let's be honest, though," Lenore flipped a whiteboard pen and popped the cap. "There is a lot we don't know yet. Automating the harvesting step in plant factories is still in the early stages of development, but having robots pick the greens will be essential."

Evan grabbed his own pen and started fidgeting. These were pointless gestures, of course, given we were talking on Zoom, but felt right somehow — after all, no matter what we may sometimes believe, we humans are analogue creatures living in a digital world.

"We will need to grow a wider variety of crops on Mars than is currently produced using these technologies on Earth. And the best cultivars possible. That means genomics. Some of that work can only happen once we are there, in Martian gravity and under Martian radiation conditions."

"Right. And securing enough water, electricity, and nutrients will be difficult."

Evan jotted down a few more points.

"Don't forget pollination. I'm not sure bees could even live on Mars, and they aren't the only pollinator we need to worry about."

"Elon Musk better have a whole team of experts planning Martian farms if he wants to avoid community failure. And failure looks ugly. Starvation, desperation, breakdown of order, all that Franklin expedition stuff we need to avoid."

We both paused for a moment, the magnitude of the problem sobering. No one wants Mars to become the site of a modern-day Jamestown.[14]

Lenore leaned back.

"I think this brings us back to where we started this chapter. If we ever do ask people to live on Mars, they will need all the extra oxygen and humidity regulation that plants provide and a place to produce small amounts of luxury crops such as spices, and they'll need green spaces just to hang out in.

"Martians will need all these things because (and we really need to keep reminding ourselves of this), people need plants around them for mental health."

---

14 Founded in 1607, the Jamestown settlement in the Colony of Virginia struggled to support itself. In 1609-10, eighty percent of the settlement died of starvation and disease, prompting a temporary abandonment of the site.

PETRICHOR

Lenore wears an unusual perfume based upon her favorite smell: petrichor. Petrichor is the earthy scent produced when rain falls onto dry earth. It is a smell that transports her back to rainy summer evenings in the forests of British Columbia and makes her feel at home wherever she goes.

The smell of petrichor was first explored scientifically in 1964 by a pair of Australian researchers,[15] Melbourne-based chemists Isabel Bear and Dick Thomas, while they were working for their national science agency, the Commonwealth Scientific and Industrial Research Organisation (CSIRO). They explained how the smell is created by oil exuded by plants during dry periods. The oils are absorbed into earth and rocks and are released when they are exposed to rain and mixed with geosmin — a by-product of bacteria — and occasionally ozone. All that chemistry makes Lenore stop in her tracks, look at the sky, and breathe deeply.

Earth-bound scientists can imagine the systems needed to keep Martian residents alive, but more nuanced elements of life away from Earth will help them thrive. Lenore had a friend who worked several tours at McMurdo Base in the Antarctic, and what Lenore's friend remembers most vividly is how dry the air was, and how she was constantly searching for Chapstick and misting her sinuses. Food at McMurdo was also a critical element of keeping a community sane in a difficult situation, and that will be much harder on Mars. Lenore wonders what Mars might smell like, what odor might define BaseTown when water hits the Martian regolith, but aside from plant factories and greenhouses, Mars will need to be a place where plants help to keep people happy, and good meals help to keep people sane.

---

15 Australia is a great place to experience petrichor in all of its glory.

Besides plant factories and good food, Martians will need a bit of a walk most days, and BaseTown's leaders might build green spaces for enjoying a bit of shinrin-yoku. This ancient practice, which arose in East Asia, consists of a slow stroll in the woods while breathing deeply. Shinrin-yoku is a form of aromatherapy. Forests are full of chemicals called phytoncides, which are the essential oils given off by the trees. These oils are antimicrobial, and it is thought that what is good for a tree is good for our lungs. Maybe a particularly large dome will house a winding forest path, filled with fragrant fruits and berries and a hot pool or two. While this might seem like a profligate use of resources that could be purposed to more practical ends, if humans ever want to reach for the stars, we need to recognize that the soul is driven by more than raw fuel. Our social, emotional, and spiritual needs must be met if we are to truly thrive.

Humans, by nature, are prone to biophilia, or the love of nature. The natural world makes us relaxed and calm. In a classic study of New York City conducted in the 1970s by sociologist William Whyte, cameras were set up across the city to study the mass movement of people as if we were any other pack animal. What Whyte learned was transformative; people are drawn to spaces with plants, water they can touch, places to sit alone and with others, and places seeded with food. Others have built upon Whyte's work to show people can stand small living quarters if they have views of nature. BaseTown's food services might well be situated in a space like The Jewel at Singapore's Changi Airport, which in the spirit of classical conservatories, features a rainforest built around a rain vortex that sends a waterfall pouring down seven stories.[16]

---

16 Changi is a destination in itself. Some airlines allow guests to check in as early as twenty-four hours early so they can properly experience the airport.

Plants will be one of the workhorse technologies of Mars. They will feed the Martians, filter the air and water, and provide everything from food to medicine to oxygen to building materials. The takeaway, however, is that humans can't thrive apart from their ecosystems. We need at least some plants and animals (and fungi as well) to build a civilization in space. When we go to Mars, if we are to succeed, we won't go alone.

CHAPTER 5:
Grass 2.0

GREEN FIELDS

On Earth, the carbohydrate is king. On a calorie-for-calorie basis, most of what we eat today comes from some form of grass. Whether it is a steaming bowl of noodles, a pot of rice, a toasted ciabatta, or even corn syrup in soda pop, about half a person's calories come directly from grain crops. Meanwhile, most of the animals we raise for food either graze on grasslands or are fed a diet of grass seeds (mostly corn). In terms of numbers, the crops that provide the bulk of humanity's calories, in order, are corn, rice, and wheat. All three grow in fields that cover millions of acres. Millions of acres we will not have on Mars.

Given how grass dominates our food system today, we need to think carefully about grass, and the role these grass crops play on Earth, when we plan future Martian food systems. Can we imagine production systems — such as a greenhouse under a domed Martian crater in the shadow of the extinct Ascraeus Mons volcano — that will allow us to produce wheat or corn? Or do we need to find alternatives that will substitute for all the carbohydrates, sugars, oils, and proteins that come from the grasses we already depend on?

Before we imagine blasting off to another part of the solar system with a load of wheat or barley seeds, we need to consider the processes that bring us the humble loaf of bread, rice noodles, and the ubiquitous industrial corn that rules the landscape of the U.S. Midwest. And then, perhaps, by understanding how this nondescript family of flowering plants has shaped life here on Earth, we can imagine engineering something that might work for Mars.

The story of grasses is the story of civilization. Almost everywhere that agriculture emerged, farmers started down the path leading to communities, towns, and cities through the domestication of grains.[1]

Wheat finds its origins in the Middle East and was domesticated ten thousand years ago in the Fertile Crescent. There have been many breakthroughs in this long history, but one of the first was at the dawn of agriculture when farmers opted to collect, save, and plant seeds from those individual plants that sported larger and more plentiful harvests and those where the ripening seeds remained attached to the stalk (rather than dropping off and falling into the dirt or explosively detaching from the stalk at the lightest touch). Having the seed remain on the plant made it easier for those early farmers to harvest. By selecting and planting these more convenient seeds, farming changed the nature of the plant, and so from the earliest days, we see human agricultural fingerprints in the genetics

---

[1] For what it's worth, the Andean civilizations got their start with tubers, not grains. One of Evan's previous books — *Empires of Food: Feast, Famine, and the Rise and Fall of Civilizations* — describes this.

of the natural world. We selected grains that were plump, uniform, and ready to harvest at the same time. Those "amber waves of grain" on the American prairies might look natural, but they are test subjects in humanity's longest-running agricultural experiment.

Plant breeding really exploded in the nineteenth and twentieth centuries. It started with Gregor Mendel's insights[2] about the hereditary nature of the color of beans and picked up pace in the 1930s and 40s when low-cost fertilizer arrived on the market. Normal Borlaug's work to breed "dwarf" varieties sealed the deal, and wheat is the poster child for this transition. These changes launched humanity into the Green Revolution, new ways of farming that boosted yields but also induced modern wheat farms to become so huge and capital-intensive as to be practically unrecognizable to farmers from an earlier generation.

Rice is a bit different. While there have been many advances in plant breeding for this crop, today, most of the world's rice is still produced on relatively small plots of land and consumed locally. Less than 10 percent of the world's rice harvest enters the global grain market to be traded internationally, (compared with almost all the world's wheat, which ends up being bought and sold on enormous global commodity markets). In China, India, and the other countries of Asia, rice is still labor-intensive with flooded rice paddies being planted, harvested, threshed, and winnowed mostly by hand. This is changing, of course, and there are large capital-intensive rice farms springing up from the Yangtze to the Mississippi, but most of the world's rice still comes from small farms that depend on hand labor.

Corn, or maize as most of the world refers to it, is a mixture of big and small farming techniques, so it is both very traditional or extremely industrialized depending on where you are in the world.

---

2   His research was discovered posthumously. Little attention was paid to the monk in his own lifetime.

Originally domesticated in Mesoamerica, maize is the backbone of the food system in South America, North America, and Africa. In South America and Africa, maize is commonly produced on small farms that use hand labor and is eaten locally — tortillas in Latin America; porridge or dumplings called nsima, fufu, and pap in Africa.

In many parts of the world where small-scale farms still dominate, farmers generally plant varieties of maize that are "unimproved" — they have not been the subject of intensive plant breeding programs. Partly because of this, many of these farmers only reap 10 to 50 percent of the yields they theoretically could if they had access to better quality seeds. Closing this yield gap is a major priority for a lot of governments and development agencies.

In the EU, Australia, the U.S., and Canada, however, corn is produced on enormous capital-intensive farms, where most of the seeds are hugely productive hybrids that need equally huge amounts of fertilizer and depend on pesticides and irrigation. In Canada and the U.S., most of the corn has also been genetically modified to make it impervious to herbicides, thus allowing farmers to spray their fields for weeds without damaging their crops. With these technologies, corn harvests in rich countries are enormous, but most of this corn isn't eaten directly by people as food. Rather, North American and European corn goes for one of four purposes: livestock feed, corn syrup, corn oil, and bioethanol we pump into our cars. The impact of this style of agriculture is enormous and immediately recognizable in satellite photos. Seen from orbit, the corn belt of North America is a distinctive green, and almost nothing aside from corn and soy is grown in this region anymore.

To the proto-Mesopotamian farmer, scratching out a meager existence with traditional varieties of corn at the start of humanity's agricultural journey, the modern corn plant would seem otherworldly. A modern corncob is about ten times larger than its ancestor wild teosinte. In addition, teosinte is a bushy plant with multiple branches, each of which sport many tiny cobs. But millennia of selective plant breeding turned this crop into something that would be unrecognizable to a farmer from ten thousand years ago.[3]

The point of this is to highlight that not only does a huge amount of our food depend on different species of grass, but that the history of farming, which is a history as large and complex as civilization itself, has been a continuous story of human ingenuity shaping the nature of nature. Everything from the genetics of the plants we grow to the way we form landscapes to support our food systems, the Earth's environment, and the grass we all depend on is a product of human intervention. All this work feeds pretty much everyone alive today, but it is also destroying our environment.

Here again, planning for the Mars mission could be extremely useful. On Mars, there are no vast tracts of land where plants naturally flourish, no great savannas or prairies waiting to be plowed. On Mars, the best we could hope to do is to put a thick transparent roof over some craters and work like hell to get enough organic matter out of the cyanobacteria biorefineries described earlier to produce maybe — at most — a few tens of square kilometers of such crops. We could never match the situation that has developed on Earth, where over seven million square kilometers of cropland are devoted to cereals. In short, the trouble with grains for the Martian is that although they produce the bulk of our calories, they simply take up too much space, hence imagining a replacement poses a real challenge.

As Evan emerged from the rabbit hole of the history of plant breeding, he texted Lenore:

"Do you think barley under glass would be doable on Mars?"

---

3   It's worth checking out the illustrations in the following paper to get a sense of the scale of how much we have transformed what we now call corn: https://www.pnas.org/content/116/12/5643

Lenore texted back immediately with a thumbs-up emoji. She knew of one place on Earth that might just be a template.

THE LAVA FIELDS OF ICELAND

It may not be Mars, but Icelandic lava fields have an other-worldly quality to them. Devoid of vegetation except a thin skein of moss and lichen, with a line of mountains in the background, Iceland is such a good stand-in for an early phase of a terra-forming experiment that bits of the BBC TV show *The Planets* were filmed there.

Lenore had, in the past, traveled to Iceland in the dead of winter. She confirmed it had a strong Martian vibe. She also knew Iceland might hold an answer to Evan's question.

Iceland is far too far north to have enough sunlight to warrant anything like a growing season, and it has little in the way of fertile soil. The landscape itself is dominated by volcanoes and lava fields. But there's something more going on in the country, just beneath the surface, that makes it a likely place to dream about how a future Martian city may sustain itself. And that something is the possibility that Mars — like Iceland — may have some geothermal heat trapped beneath the surface. And all this extra heat has allowed Icelandic people to develop an outstanding greenhouse industry.

Lenore had seen some of that technology at work: hydrothermal energy harnessed to produce fresh greens, tomatoes, cucumbers, and even exotics like avocadoes. Iceland's greenhouses are like beacons of warmth and light in the near total-darkness of winter.

Could the same technology work on Mars? Although the volcanoes of Mars seem to be extinct, scientists think that the planet may still have a molten core. Because of this, Martian enthusiasts hope there may be geological anomalies that could provide heat and energy close enough to the surface for a future community to tap.[4]

Evan had every intention of visiting one of these Icelandic greenhouses to research this book, determined to treat himself to some science tourism. But the world had other ideas; he found himself spending embarrassing amounts of time on YouTube staring at Icelandic landscapes and forcing his wife to watch endless murder mysteries set in Scandinavia.

It was during one of these sessions that Evan stumbled across one of the most fascinating research projects he uncovered in the entire year of research. It was a project that involves crazy, cutting-edge science, gene splicing, and geothermal-fueled Icelandic greenhouses, all to grow grass close to the Arctic Circle.

Like so many things in the modern age, Evan's discovery of this science experiment began with a Google search: grass + greenhouse + Iceland. It wasn't long before he was staring, somewhat open-mouthed, at the computer screen, watching a lecture by Dr. Björn Örvar, the co-founder and chief scientist of ORF Genetics (who incidentally received his Ph.D. from Canada's own UBC). The video was shot in a greenhouse where Dr. Örvar was flanked by rows of barley plants growing in pots and trays, the Icelandic volcanoes in the background.

---

4   In fact, some of the recent Mars missions are pouring cold water on this hope, and it increasingly seems unlikely that there will be major geothermal energy sources found on Mars. Likely, Martian communities will need to rely on nuclear reactors and possibly space mirrors (to concentrate the sun's energy). However, if some accessible geothermal sources are discovered, then it will make the prospects of the future community much more realistic.

Warmed with geothermal energy, and lit with electricity-powered geothermal generators, the ORF Genetics generators, the ORF Genetics greenhouses grow greenhouses grow genetically modified barley that is engineered to produce proteins normally only found in mammals. The company is the brainchild of Dr. Örvar and three other scientists whose first project was to trick their barley into producing a series of proteins found only in human skin. While this may sound preposterous, they were quickly able to get this ingredient okayed for the cosmetics industry and today one-quarter of Icelandic women above the age of thirty use OFR's patented anti-aging skin cream.

More recently, the team at ORF has bred barley that creates proteins useful in both stem cell research and — more importantly for this book — for companies growing meat in petri dishes (cellular agriculture).

ORF Genetics uses their gene-editing expertise to reprogram their barley plants to produce what scientists call growth factors, proteins that help other cells grow and divide. Put simply, if you are doing work that uses specific cell lines (such as would be used in cellular agriculture), and you want to create more of these cells, you can add ORF's barley-mammal-protein to the mix to speed up how fast your cells divide. So, ORF grows and harvests their fancy barley seeds in the Icelandic greenhouses, extracts the proteins responsible for the growth factors, and ships these growth factors to cellular agriculture start-ups who use them to accelerate growth in their bioreactors, a topic we will explore later.

Hands in the pockets of his white lab coat, ruffled salt-and-pepper hair, and sporting a five-o'clock shadow, Dr. Örvar carefully explains to the camera ORF's approach and why they use barley.

"[Barley] is one of the oldest and most robust agricultural plants. And it has high climactic adaptability compared with other agricultural plants."

He then points out that barley can self-pollinate, making it easier to control how the genetics from one generation are transferred to the next. Because of this, ORF can "drastically reduce the cost of producing these growth factors for cell-cultured meat."

Dr. Örvar finishes his lecture on an aspirational note.

"Our mission is to remove the biggest roadblock cell-cultured meat producers are facing. To feed the world's growing demand for cruelty-free climate-friendly meat, the cell-cultured meat industry needs a solution." And to him, the solution is to use their barley-derived growth factors to help make cultured meat more efficient.

"So basically, what we've got," explained Evan to Lenore later, "is a company that grows genetically modified barley on a landscape that looks like Mars. And it gets better. Their barley actually creates a protein that helps speed up the bioreactors that create cell-meat."

Lenore paused. "So, they aren't using their barley to produce bread . . . or even beer . . . ?" Her tone suggested she was getting impatient — after all, this is supposed to be about food.

"True. But remember, most of the grass we grow in North America is industrial corn that is fed to pigs, chickens, or cows. North Americans really don't eat much corn, other than the corn syrup in soda, so when you peel back the layers, the industrial food system on Earth today is geared at using genetically modified grasses that are fed to other organisms as a way of creating protein that people like to eat. I'd say what ORF is doing is pretty much the same. Only they are replacing pigs, chickens, and cows with microorganisms that digests starches and sugars to produce faux pig, chicken, or cow protein."

But Lenore still wasn't totally convinced.

"Maybe, but I'm not sure. To be honest, the idea of genetically modified barley, growing it with geothermal heat, near the Arctic

Circle, all to create an input that makes test-tube meat grow faster feels a bit dystopian … like the plot of a bad sci-fi with mad scientists running amok?"

She trailed off.

"It's all connected," said Evan, the optimistic voice for once. "ORF's greenhouses in Iceland are simply a recent example of our species' ten-thousand-year history of modifying the genetics of the *Poaceae* family to solve problems in the food system and keep us all fed."

"Plus," he continued, "it's possible that in the future they might even be able to separate out the sugars and starches from the proteins in the barley and then recycle the rest of the plant so that a single barley crop could produce these very valuable proteins, possibly be a source of some starch or sugar that could be used for other things like brewing, and the rest of the plant would end up as compost. On Mars, every organic molecule will be a resource, so the folks who end up on that first community will be looking to find many possible uses for absolutely everything."

Lenore nodded. "It's a start, I suppose. But we eat a lot of carbohydrates. Won't anyone think of the pasta? Where are we going to source those calories?"

THE PROBLEMS WITH GRASS

Maybe part of the answer is we don't need all those carbohydrates. They are mostly a food of opportunity — easy to produce, so we produce them, and then consume them. And besides, on Earth we do not have anything like the kind of efficient system Evan has been imagining will be imperative on Mars. Perhaps Mother Earth has historically been so fertile, so abundant in topsoil and verdant plains, that we've been able to be lax about resources and have let our food systems evolve into a series of systems that are very ineffi-cient in terms of inputs.

For all our successes in breeding productive varieties of corn, rice, wheat, and barley (and making delicious beverages along with tasty bread and noodles out of the resulting crops), humanity has a major problem with grass. Put simply, the way we produce and use the stuff is driving some of the most serious human and planetary health problems around, not to mention driving the planet's wild-life into a marginal situation . . . and all of this is partly thanks to its very abundance.

Let's start with the environment. Pastureland and grain fields have the biggest footprint of any human activity, and our pastures (which are just semi-wild managed grasslands) alone take up more land than all other human land uses combined. As such, growing grass for crops or grazing is a key factor driving our losing fight to save the world's biodiversity. Equally detrimental, all this grass causes a huge amount of greenhouse gases to be emitted. About one-quarter of the world's GHGs come from growing grain, the livestock that eats the grain (such as the methane produced by the world's 1.5 billion cows [5]), or the energy used to transport the resulting food. If these numbers seem too large to make much sense, then perhaps a metaphor is more apt: the collective impact of grain fields, pastures, and livestock is akin to a slow-moving asteroid strike.

But the corollary of this is also true. And if today's global grain/grass/livestock systems represent a massive environmental problem,

---

5 According to the UN's Food and Agriculture Organization, in 2019 there were 1,511,021,075 cows on the planet. Other data sources have similar estimates: between 1 and 1.5 billion at any given time over the last five years. If you use a round figure and assume that each cow weighs about a thousand pounds, as compared with the average human who is about 140 pounds, this means that on a pound-per-pound basis, there is probably more cow biomass on the planet than human biomass (and there are a lot of people).

then this also means that some of the best ways of saving our environment come from the transformation of this very system. By working to better understand soils and the plant microbiome, by investing in technologies that will make farming more precise with inputs, and by switching to more alternative proteins, our food systems can evolve and adapt. And the systems we are designing for Mars will help point the way.

There are also a host of human health impacts caused by eating all this grass. Starting first in Western countries, but now spreading all over the world, for the last two generations, doctors, nurses, and dietitians have observed skyrocketing rates of diabetes, obesity, and other chronic diseases linked with diet. One cause is, of course, the fact we now eat excessive amounts of empty carbohydrate calories, often in the form of refined grains and refined sugars, most of which come from corn, wheat, or rice. And it's also likely that all these refined products are changing the bacterial communities that live in our internal microbiomes, and so eating too many refined grains is implicated in what seems to be a rise in intestinal problems like celiac disease. Industrial corn production — corn is one of North America's biggest crops — is a particular problem as it is the basis of high fructose corn syrup and cornstarch, key ingredients in pop, chips, and candy. Nutritional scientists talk about the "dietary transition" as citizens in more and more countries embrace the wonders of the North American diet and gorge themselves on junk food.

To put the scale of this nutritional problem into context, a couple of years ago, Evan worked with a research team that compared dietary recommendations with global agricultural production data.[6]

---

6   KC K., Dias G., Veeramani, A., Swanton C., Fraser D., Steinke D., et al. (2018) When too much isn't enough: Does current food production meet global nutritional needs? *PLoS ONE* 13(10): e0205683. https://doi.org/10.1371/journal. pone.0205683

Although the paper is a few years old, he still gets excited about this research that compared what we *should* be eating with data on what we *are* producing. We know that about half of our diets *should* be fruits and vegetables, but the world only produces a fraction of this, so we don't have anywhere near what we would need if everyone ate the diet nutritionists recommend.

Evan and the team of researchers started with dietary recommendations such as the Canadian Food Guide and Harvard University's Healthy Eating Plate model, both of which show fruits and vegetables should be about half of what we eat. The researchers then calculated that today fruits and vegetables are only about one-eighth of the total food we produce. Similarly, although nutritionists suggest that about a quarter of our diet should be grains — and ideally these should be whole grains — close to 50 percent of the world's food supply are grains that are mostly eaten in some sort of refined form. Plus, we overproduce fats and sugar. The conclusion of this research was that there is a fundamental mismatch between what we know we should eat to be healthy and what we are producing in the world's agrifood systems.

On Mars, the food system must be optimized for nutrition. Grasses might play much less of a role than on Earth. Though Evan and Lenore first met over donuts, the Martian diet will be light on sweets and pastries and heavy on fresh strawberries and artful salads. A similar model isn't necessary on our home planet, but the truth remains: terrestrial agriculture is harming our personal health and destroying the planet, partly because of the quantity of grains and livestock products we consume. Feeding Mars is a chance to build a food system from first principles divorced from the mistakes of history.

There are a lot of reasons why both the environmental and the health problems linked with humanity's unhealthy addiction to grass have emerged. Partly, it seems to be down to our genetics: many people really enjoy the taste of carbs and sugars. There may well be evolutionary origins to these cravings, and it doesn't take much imagination to consider that during an earlier stage of human history, those folks who were highly motivated to obtain sugars were also more successful in getting themselves and their children through the colder stretches of the Ice Ages. Today, of course, we are awash in carbs and sugars and so those cravings may have the opposite effect and hurt our life expectancy. Lenore has never seen a slice of pie she didn't like, and Evan loves his donuts, but both of us should be walking past those sugar bombs to get to the salad bar.

But an evolutionary predisposition to a sweet/carb tooth goes only part of the way to explaining why farmers produce so much grass and not enough fruit and veg. Some of the overproduction of grains is due to policy. For decades, many countries have subsidized producing corn. The U.S. has been one of the worst culprits, pouring billions of dollars per year into the pockets of corn producers and flooding the market with taxpayer-subsidized grain. While this has changed somewhat in recent years, the bottom line is that around the world, government subsidies have contributed to the overproduction of grain crops.

Governments have also poured money into researching grain crops. In fact, most of the world's agricultural funding for the last century has gone into boosting rice, wheat, and corn production. By contrast, the investment in fruit and vegetable research is pennies on the dollar when compared with what we've directed to commodities and grains. Historically, grains kept people (and armies) fed. Today, we struggle with a glut of carbohydrates.

This, in a way, is sort of a good news story. If a lot of the production of grains is a result of subsidy and policy, then maybe we can assume that Martians will eat something closer to a nutritionally "recommended" diet rather than indulging in what is today the average Earthling's diet. If this assumption is correct, then the need to produce sugars, carbohydrates, and proteins (and livestock) is greatly reduced. But even still, there will be a need for grasses on Mars as no one expects a Martian to live by algae alone. After all, there needs to be at least one donut shop in BaseTown's main mezzanine. Where will those grain crops come from?

HACKING PHOTOSYNTHESIS

After working intensively on this chapter for a few weeks, Evan was feeling a bit lost in the complexities of dietary guidelines, the sustainability of different types of farming systems, and the sheer improbability of imagining Martian wheat fields. He called Lenore, and they took things back to the basics.

"OK," summarized Lenore, "I think that there are three key points you are circling around, three points that both capture what you've told me about grass and that are applicable to Martian and Earthly food systems.

"The first is that, while we'll likely need to produce some grain on Mars, it takes up too much space to grow, so grains will only be a small part of a Martian's diet."

Evan nodded, "Yup, 100 percent."

"Second," she continued, "maybe only having a small amount of grain on Mars isn't too big a deal as we are already eating too much of it, so we should be cutting back on grains anyway."

Evan nodded.

"Third, we could also do sophisticated breeding and genetic work so that when we take grass to Mars, it is genetically suited to a Martian habitat and engineered to produce nutrition along with other useful proteins, much like ORF Genetics in Iceland is using barley to produce growth factors and pharmaceuticals.

"Basically, if we can do these three things — produce a small amount of grain, eat a more balanced diet, and make sure you get the genetics of the grass optimized for Mars, then I'm guessing it will make sense to invest the time, energy, and resources to take grass to outer space."

"Yes," Evan said, "I think so, and this also reminds me of something."[7]

Several years ago, Evan sat down with a friend over drinks before a conference on food security. This friend worked for the Bill and Melinda Gates Foundation, and Evan asked his colleague where he thought the next big breakthroughs in agricultural research were going to come from. After all, the Gates Foundation has a reputation (and received a fair bit of criticism) for going after big technological moonshots to some of the world's most complex problems, so who better to read the techno tea leaves than a senior person there?

The reply to Evan's question came quickly and confidently: hacking photosynthesis. In particular, the goal would be to use genetic engineering to make the molecular basis of plant life more efficient. This was described as a high-risk/high-reward project that was scientifically challenging but could be transformative.

The logic behind hacking photosynthesis is relatively simple. As discussed earlier, over the past ten thousand years farmers, plant breeders, and scientists have changed the crops we depend on to such an extent that many of our key crops would be practically unrecognizable to early farmers. Yields have exploded, keeping global food production comfortably ahead of population growth,

---

7   This anecdote is described in some detail in chapter 7 of Evan's 2020 co-authored book *Uncertain Harvest*.

and today, there are over 2,800 dietary calories produced for every man, woman, and child.

But despite our tremendous successes in breeding varieties of grain that are extremely productive, there is one key variable that we have not really nudged the dial on. In all this effort, the efficiency of photosynthesis hasn't changed at all.

Photosynthetic efficiency is defined as the ratio of the amount of energy a plant creates and stores (generally as sugars) versus the amount of the sun's energy that falls on that plant's leaves. Now there is a lot of complex science and math that goes into calculating photosynthetic efficiency,[8] but most of the research suggests that plants can turn only 3 to 6 percent of the total solar radiation that lands on their leaves into sugars (aka, chemical energy).

Figuring out how to increase the amount of the sun's energy a plant can use is a critical scientific goal that might make both Earthly and Martian food systems much more viable. On Mars, much farther from the sun, there is a relative lack of solar energy.[9]

---

8   The exact number you get when calculating photosynthetic efficiency depends on how you define plant energy and how much of the solar spectrum you consider to be light. The numbers quoted here seem to be the most commonly used in the scientific literature.

9   Pinning down an exact set of numbers to compare the amount of solar energy on the two planets is tough. For one thing, both have slightly elliptical orbits, which means that both are at different distances from the sun at different times of the year. Dust storms on Mars block the sun for long periods of time. The angle of the sun is also different on each planet and at different times of the day and year. Hence, scientists talk about *solar insolation*, which is the actual amount of solar radiation that arrives on a horizonal space over a given period of time. The bottom line is that Mars generally receives somewhere around 50 percent of the solar insolation of Earth. This means that putting a greenhouse at the Martian equator would be somewhat analogous to putting a greenhouse on Devon Island in Northern Canada.

So, having plants that can be more efficient with the sun's energy could be extremely important.

But trying to boost photosynthetic efficiency is needed on Earth too. Here the demand for food is rising with population growth and rising incomes (mostly in Asia) that are lifting people out of poverty. Add to this the fact that climate change and water scarcity may make many of our planet's most productive regions unproductive, and the inevitable conclusion is that we are going to need to boost production in those parts of the planet that can sustain crop growth. But cropland is already maxed out, so unless we want to pillage the remaining forests of the world, we need to grow food more intensively than we currently do.

Boosting fruit and vegetable production without expanding cropland can be done in the sort of indoor systems described in chapter 4. But recent research suggests that hacking into the genetics of photosynthesis could help grain farmers in the future produce more of these commodities on less land while also helping these crops remain productive during extreme weather.

While the science is complex, some of this work is easy to explain, and it mostly has to do with how plants use carbon dioxide. Basically, when temperatures get hot, most plants close their stomata (pores in the leaves) to conserve moisture, but this reduces the plants' ability to absorb carbon dioxide, reducing the efficiency of photosynthesis. But a few species of plants — called C_4 plants — can keep photosynthesis going even when it gets hot. Overall, only about 3 percent of flowering plants are C_4, but together this group produces about 20 percent of global photosynthesis. So, several scientific teams — including some funded by Gates — are working to better understand how C_4 plants operate and are trying to figure out how to use these insights to engineer extra productivity into some of our key crops. While there are other approaches to trying to hack photosynthesis, working on this C_4-pathway is one of the most promising. Some recent research suggests that it may be possible to boost yields by something like 40 percent using this kind of bioengineering.[10]

Producing 40 percent more food on the same amount of land would be transformative for global food security. And tweaking the way that plants convert carbon dioxide into sugars could make grain production possible for a Martian community. But at this point, Evan decided he needed to come clean to Lenore on an issue that was really starting to bother him.

THE MARTIAN MINDSET

"OK," Evan began, leaving dreams of Mars aside for a moment, "full disclosure. A couple years ago, I was working on an academic book about climate change and food systems with two brilliant emerging scholars named Ian Mosby and Sarah Rotz. In that book, we looked at all this Gates-funded work on hacking photosynthesis and spent a large amount of time debating whether we thought this technology was really going to deliver the hoped-for breakthroughs.

"After a lot of discussion, we decided that maybe the rhetoric around DNA hacking was a bit overblown and this whole topic seemed overhyped. A lot of consumers and regulatory bodies like

---

10 There is a lot published on this topic. Here is a brief sample: (1) Carl R. Woese Institute for Genomic Biology, University of Illinois at Urbana-Champaign. (2020, August 10). Photosynthetic hacks can boost crop yield, conserve water. ScienceDaily. Retrieved January 29, 2022, from www.sciencedaily.com/releases/2020/08/200810113213.htm (2) South, P.F., Cavanagh, A.P., Liu, H.W., & Ort, D.R. (2019). Synthetic glycolate metabolism pathways stimulate crop growth and productivity in the field. *Science*, 363(6422). (3) Sabtora, T. (2019, October 28). Hacking photosynthesis to feed the future. Scienceline. Retrieved January 29, 2022, from https://scienceline.org/2019/07/hacking-photosynthesis-to-feed-the-future/

the EU hate genetic modification, and a lot of folks think it would be better to try to use more natural plant breeding approaches to try to get the same outcomes. So, at the time, photosynthesis hacking felt like the 'mother of all techno-fixes.' We concluded that simply enacting good public policy (such as putting a solid price on greenhouse gas emissions or funding better plant breeding science) would be a more streamlined and equitable way of getting us a more sustainable food system, rather than pinning hopes on a big scientific breakthrough.

"But I'm now having a crisis of faith. Since Ian, Sarah, and I worked on that book, which was just a couple years back, the world has lost between 400 and 4,000 species[11] and emitted somewhere around 90 billion tons of carbon dioxide. Meanwhile, the planet is now about 0.25 degrees Fahrenheit hotter and the human population has grown by about 160 million souls. In 2019, 1.8 million hectares of forest burned down in Canada alone.

"So, things have become significantly worse for planet Earth while we've been waiting for good policy.

"Over the same period, some of the top teams of researchers funded by Gates to work on photosynthesis hacking have continued to publish impressive papers showing that they are knocking off some of the major barriers standing in the way of engineering much more efficient crops. Right now, it feels like we are falling behind, that policy isn't really helping, and that science and technology are creating real solutions.

"Let me tell you about a couple of young scientist-entrepreneurs I've met recently. This pair — Luke Young and Rory Hornby — have started a venture called Agrisea where they have identified just seven genes in the rice genome that they think are responsible

---

11   A fact Lenore knew only too well, having written a book on extinction.

for salt tolerance. Basically, Luke and Rory are developing a variety of rice that should be able to tolerate salty conditions and may even grow in brackish water. Thanks to unsustainable farming and climate change, small-scale farmers across much of Southeast Asia are struggling to keep producing as their paddies get inundated with salt water, so developing salt-tolerant rice could be huge. It could even lead to producing rice on floating islands in coastal regions. And if we think of Mars, it's possible that Agrisea's genomic work could lead us to some basic grain crops suitable for Mars.

"I still believe in good public policy. But we're running out of time, and it seems that the scientists are doing a better job than the politicians — is this the horse we should back?"

There was a moment of silence as Lenore digested this. "Science works," came her slow reply. "But it's not a panacea — it never is — but it does seem to work. Meanwhile, politics is slow. And I agree. We are running out of time. It's one of the reasons projects like building a community on Mars could be valuable. Such projects supercharge lines of research that sometimes get mired in politics and policy on Earth. You know how I feel about the unreasonable fear of GMOs. Or fear of anything, really. Evan, only well-fed people have the luxury of being all NIMBY about technology."

We kept debating this point. In the endless iterations of this conversation, Lenore usually comes down harder on the potential for science and technology to save the day. Evan is usually a bit more cautious and desperately thinks that underpinning good science and technology needs to be a foundation of strong progressive policy. But we agree that both science and policy are needed. And we agree that humanity is running out of time.

So, to draw this chapter to a close, at the time of writing, here is our reading of the tea leaves about photosynthesis hacking.

First, we think that by the 2030s, scientists will almost certainly have achieved some major breakthroughs in their ability to understand, and engineer, plant genetics such that future farmers will have access to seeds that remain productive under hot, dry conditions and produce much more food on existing cropland.

Whether these scientific accomplishments result in a better-managed environment, or a healthier human population, will depend on a lot of social and economic factors. For instance, we will need to determine if the resulting high-productivity seeds will be made available to farmers in the Global South either free or at a cost they can afford. Determining the answer to this question — which is a matter for public policy — will determine the use of this science.

Second, it will be a matter for public policy and consumer behavior whether this science simply is used to let us continue churning out huge amounts of cheap corn syrup or whether this science results in healthier food for all.

Third, one of the hoped-for benefits of these technologies is to increase the amount of food that is produced on current farmland and that this will reduce our need to press further into the world's forests. But this is hypothetical: the idea that farming intensively in one area will spare land somewhere else is only a theory. Whether this happens, whether nature is saved, is a matter for policy makers and planners. If we want land protected, having wonderful seeds to boost yields is only one part of the solution. Strong policy to protect nature is equally (if not more) important. In other words, science provides tools, but social, political, and economic factors determine whether these tools become solutions.

When it comes to Mars, in some ways, our thought experiment is simpler. We think that one potentially viable strategy to provide BaseTown with a limited supply of grains and oilseeds would be to use space mirrors in orbit around Mars to focus the sun's energy onto a small number of Martian craters that would have to be covered with heavily reinforced and tinted glass that would filter out the harmful parts of the radiation (it's possible that ice could also be used as a building material for this). Inside these giant greenhouses, farmers would grow grain on organic substrates obtained from the cyanobacteria tanks and the composting facilities. The plants themselves would be inoculated with beneficial soil microorganisms that would — among other things — fix nitrogen from the atmosphere. Grain plants would be specially bred and genetically engineered to survive in the low gravity and low pressure of Mars and also be engineered to produce medicinal proteins along with starches, carbohydrates, oils, and sugars.

At harvest, each of these constituent parts would be carefully separated, with the highest-value products (likely the pharmaceuticals) being taken first, before other components are turned into cooking oil, malting sugars, or flour. Everything else would be recycled immediately back into the food system via a composting infrastructure that would collect both the organics and water. Composting is also a source of heat, which would be trapped and used to generate power. Overall, this would mean that the folks of BaseTown would enjoy a modest amount of grain products in their diet, but ingredients such as flour or corn syrup would be a luxury and saved for special occasions. Pecan pie, one of Lenore's favorites, would be a food for special occasions, a hardship balanced by an unlimited supply of fresh mangos growing here and there around the public areas of town.

"You know," said Lenore, "one of the key lessons I'm learning may be more about the mindset of being a Martian than the specifics of any particular technology. To imagine a Martian food system, we are forcing ourselves to think about how every single input can be used with total efficiency, and that every output is imagined fulfilling multiple purposes. Maybe it's this mindset — more than the technologies themselves — that we need to adopt on Earth?"

With that, Lenore headed off to go down the deepest, most difficult rabbit hole of them all. One of the most wasteful parts of the food system on Earth is the production of animal protein. Would Martians be vegan? Could Mars ever be a home where the buffalo roam?

PART III:

RED MEAT

CHAPTER 6:
I Can't Believe It's Not Cow

INSPIRING MARTIAN PLANNERS WITH FIKAS AND FJORDS

Let's take a tiny stroll back in time, not too far back, but before the pandemic when Lenore was still doing what she does best, what she was born to do: wandering the world, eating things. And in this flashback, she was wandering the streets of Stockholm.

If future Martian civic planners are feeling bold, they might look to Stockholm for inspiration. Fast and efficient public transit, clean and pedestrian-friendly streets, and the renowned Scandinavian social safety net lie just under the surface as healthy, happy citizens go about their business of leading relaxed and balanced lives. And — once or twice a day — these same citizens emerge from neat midrise stone buildings and take to the cafes for fika, the practice of stopping to enjoy conversation, strong coffee, and something sweet.

Lenore was in Stockholm to immerse herself in fika culture as part of a quest to understand the history and evolution of the modern cinnamon roll. Many fika breaks feature a kanelbullar, a classic rolled pastry that is beloved by modern Swedes, who consume an average of 316 buns per person per year. Unlike the heavily frosted monstrosities served up in North America, the Scandinavian bun is a more restrained affair — a delicately kneaded spiraled dough that makes a pinwheel of about the same diameter as the coffee cup that they accompany. But what a kanelbullar lacks in size it makes up in flavor, dominated by a heavy spicing of cinnamon and cardamom. Swedes, in fact, consume eighteen times more cardamom per capita than the average country and have done so for over one thousand years, since the Vikings brought the spice back from the Moorish peoples of Spain. This unusual, but tasty, habit is one of many medieval twists present in the Scandinavian countries' cuisines.

Lenore is a fika-finder extraordinaire and ended up sitting at a delightful little sidewalk café in a waterfront park watching well-behaved children flit about as the sun dappled the grand palace on the central isle of Gamlastan. Lenore was already thinking about how future food systems might evolve, and she was already chatting with Evan, though Mars was still over their horizons. A question danced on her mind, and she whipped out her phone and texted him.

"What food product in Northern Europe do you think is steeped in culture and has the most problematic carbon footprint?"

Evan's one-word reply came back immediately: "Milk."

Sweden is a milk-loving nation. The country ranks first or second globally in dairy consumption per capita, and though only about 6 percent of the country can be farmed, about a third of that is pasture. Dairy is the top agricultural product in Sweden with about 3,500 dairy farms (though that number is falling). On the streets of Stockholm, yogurt and cheese are everywhere. On Earth or Mars, expat Swedes are going to want their ost (cheese).

But doing dairy on Mars will be a challenge. Even though some of the earliest colonists to North America were bovine, cows generally don't travel well. It's possible that we could take cow embryos along on the ride between Earth and Mars, but the radiation exposure would be terrible for the fetal creatures, and then there is the question of what on Mars the animals would eat when they got there. After all, we are imagining grass will be in short supply.

In addition, we know shockingly little about how animals fare in space. To date, seven nations have sent animals into orbit for brief experiments in zero-G: the Soviet Union, the United States, Argentina, France, China, Iran, and Japan. But to date, no major agricultural animals have made the list (the International Space Station once hosted a quail), and the only animals of any size to reach orbit have been a few dozen Soviet dogs, a single cat, and a collection of monkeys. These unfortunate and unwilling participants in our space program include Laika, a space dog of the Soviet Union; Ham (for Holloman Aerospace Medical Centre), a chimpanzee used to test how people might respond to high-G blastoff; and a host of other lost, nameless souls. The results of these tests, though spotty, don't suggest much good news for would-be space ranchers.

Like humans, animals in space experience stress and lose weight, muscle mass, and bone density. But even if we could move a few cows to Mars, we likely won't. A dairy cow eats twenty-five kilograms of feed a day and drinks between 100 and 200 liters of water. On Earth, feeding these creatures represents about 70 percent of the operating cost of most dairies. In a pioneering space flight to Mars, the energy required to take such vast quantities of food and drink — to say nothing of the animal itself — off Earth would crush even the most generous budget. And producing so much food each day just to extract a bit of milk from a large quadruped is utterly impractical. So, the idea that we will ever hear the gentle mooing of *Bos taurus* as the sun crests on a Martian dawn exceeds even the most fanciful science fiction.

Fortunately for our hypothetical Martians, there are other options. This is also fortunate for Lenore; she is part of the majority of *Homo sapiens* who have some degree of intolerance to dairy. This developed in early adulthood; one of those all-time favorite childhood treats — the milkshake — now gives Lenore what can politely be called a period of uncomfortable unpleasantness (Evan, by contrast, probably has Viking genes and can guzzle cream with abandon). Milk contains a sugar called lactose, and when we are children, our bodies produce an enzyme called lactase that allows us to digest milk. For many of us, our ability to produce this enzyme fades with time, and as it goes, so, too, does our ability to digest milk and other dairy. Globally, about two-thirds of the human population react to milk in this way, though some populations can stomach dairy better than others. The mutation allowing us to drink milk in adulthood is called *lactase persistence*. In Northern Europe, where for millennia people have relied on dairy products to survive harsh winters, 90 percent of people carry this mutation, but in Asia and South America, the rate is much lower. The mutation appeared about 4,300 years ago, likely marching arm in arm with an explosive expansion of dairy agriculture and perhaps even spread from Scandinavia to Scotland (where Evan's family originally comes from) through Viking settlement.

So, what does a food professor without lactase persistence do on a food tour of dairy-rich Sweden? Well, it is possible to take lactase tablets and give your intestines a short-term boost just before you gorge yourself on ost. But in many ways, the best option has always been to use plants as substitutes for dairy. Replacing milk with plant-derived products — such as almond or soy "milk" — is old technology. The first such product was likely from soy and appeared in China about two thousand years ago. Similar beverages include the horchata of Central America, the boza of Eastern Europe, and the malted millet beverages of East Africa. Most of these alternative milks are traditionally made at home in small batches, but in the 1940s, a commercial soymilk industry developed in Hong Kong, spreading globally in the 1970s and 1980s.

The earliest versions of this product were a bit grim and are remembered by Evan, Lenore, and many other children of the 1970s with a shudder for their gritty texture and "beany" flavor. The first attempts at soy ice cream were both brittle and bitter (not two words usually associated with a nice trip to the ice cream parlor). Since then, however, plant-based "milk" technology has advanced markedly and today it is driven by the popularity of plant-based diets, health concerns, worries over the environment, and — of course — lactose intolerance. Soy milk continues to dominate the market, but cashew milk, rice milk, coconut milk, and even macadamia nut milk all have strong fan bases. In the U.S. alone, the plant-based milk category is worth about two billion dollars per year — or 14 percent of the total retail milk market.

But before we blast back over to Mars, both Evan and Lenore think that the real story in the plant-based milk community is the new kid on the block: oat milk.

DAVID, GOLIATH, AND THE SWEDISH MILK INDUSTRY

Back in Stockholm, Lenore was about to have her first encounter with the phenomenon known as Oatly. She had vaguely heard of this cult brand of oat milk as it had received a lot of fan press in North America.

Lenore didn't seek Oatly out at first. She was scarred by her early encounters with the 1.0 versions of North American oat milk that were . . . not good. But as she prowled her first Swedish grocery store, she found herself facing a wall of brightly colored packages proudly bearing the logo of the oat milk that trendy baristas covet.

She picked up a box. On the side was a jaunty tale of happy farmers, energetic entrepreneurs, healthy guts, and a well-managed environment. Maybe it was the color, maybe it was the oddly engaging stories, but the marketing was good, and after grabbing a few different products, Lenore did some tasting. And here is the thing; as a pair of culinary geographers, both of us think the truth is undeniable. Oatly is a great product.

Lenore headed out immediately to find an Oatly milkshake.

"Licorice?"

"Ja. Dagens smak."

So, the daily oat milkshake flavor was licorice.[1]

"Hmm . . . ?"

She texted Evan who paused marking essays to glance at his phone: "Would you drink an oat milk licorice milkshake?"

"Hell yeah!" came his instant reply, complete with a high-five emoji. (Truth be told, Evan was jealous of Lenore's trip and secretly hoped the milkshake sucked.)

Skeptical that licorice has any business in ice cream, Lenore asked again. The bemused server was firm — licorice was the oat-milkshake flavor of the day.

Lenore glanced at Evan's high-five emoji and took a sip. The licorice flavor was a bit of a distraction, but the mouthfeel of Oatly did a good job mimicking dairy. Nutritionally, it's basically like drinking a bowl of porridge, and unlike soy or almond milk — each of which come with serious environmental costs — growing oats gets you a gold star for sustainability. Oats are an important part of a sustainable crop rotation but have been on the decline for decades

---

1   Licorice, or *Glycyrrhiza glabra*, is a flowering perennial with a thick root that yields a sweet aromatic flavoring. Swedish people love licorice and will likely take it to Mars, where it will appreciate the sandy soil and puzzle non-Swedish milkshake drinkers.

due to a lack of market since the automobile replaced horses as our preferred way of getting around. Here, perhaps, is a milk-substitute worthy of support?

But how does one milk an oat? And what does this have to do with our hypothetical Martian community?

The answer to the first of these questions is chemistry, and the second answer is almost as simple: once we understand the chemistry of producing oat milk, then we'll have some of the tools needed to ensure that our future space-age cousins will be able to enjoy a nice plant-based latte before a hard day of terraforming.

In Oatly's case, the story of getting the milk from the plant begins when enzymes are added to oats, liquefying them but keeping the beta-glucans (a form of naturally occurring sugars) in the oats intact. This process allows Oatly to gently foam and produce a reasonable latte as well as giving oat-based ice cream a decent texture. The process of doing this was formulated by food scientist Rickard Öste at Lund University, and his company, founded in 1994, has been turning out quality oat products ever since. Demand for oat-based dairy substitutes, though, built very slowly, and the popularity of Oatly has really grown only in the last ten years. Oatly does, however, enjoy a cadre of early-adopting true believers who emerged due to a combination of barista culture and a lawsuit that backfired spectacularly.

The lawsuit story is a David meets Goliath tale that began when the giant Swedish dairy industry took Oatly to court after the plant-based start-up cheekily claimed that they were producing "milk for humans." While Oatly lost the battle, and the judge agreed with the traditional dairy industry that their marketing wasn't OK, they may have won the war. Oatly responded in the court of public opinion and published the text of the lawsuit, framing it for what it was: an age-old tale of an established industry attacking a tiny new entrant. Sales of Oatly skyrocketed, and the company expanded overseas. They have never looked back.

Oatly is widely available now, and they've grown throughout the pandemic. In 2020, dollar sales of oat milk were up 212 percent annually in the U.S., and oat milk is now the second most popular plant alternative after soy. In 2020, Oatly produced 165 million liters of product, an increase of 93 percent over the previous year. And in May 2021, Oatly's name recognition exploded in North America when its initial public offering appeared on the Nasdaq stock exchange.

Later, Lenore remembered her oatshake, and she texted Evan: "I wonder if oat milk will be the dairy of choice for the Martian on the go?"

"Maybe," he replied, but by this point, he was starting to research growing grain on Mars and getting a bit skeptical (see chapter 5). While we think that it is possible a future Martian community may be able to produce a few oats and a reasonable (but modest) supply of oat milk, this won't suit all the culinary needs of the people of BaseTown. For one thing, even oat milk struggles to make one of the most popular dairy products of all-time: cheese.

UMAMI AND TERROIR

In the world of cuisine, cheese matters. To understand why cheese is, and will continue to be, important on and off this planet, we need to understand two words that come from very different ends of the Earth: umami and terroir.

Umami comes from the Japanese and explains why the human palette is so drawn to dairy ferments such as cheese. *Terroir* comes from France and speaks to how food produced in one region can be completely different from the same food produced elsewhere.

Although not yet well recognized in the West, umami is a basic human taste like sweet and sour and is best described as a delicious and satisfyingly savory sensation. Chemically, umami is triggered by a handful of chemicals that, when eaten, light up receptors on our tongue like a pinball machine. We crave umami foods, and we miss them if we can't get them.[2]

For such a fundamental culinary building block, our understanding of umami was late in developing. Around the world, people noticed that savory foods were pleasant, but they couldn't exactly say why. The great French chef Auguste Escoffier noticed a pleasant hidden flavor in certain food combinations, but an actual understanding of umami didn't occur until 1908, when chemist Kikunae Ikeda of the Imperial University of Tokyo became obsessed with understanding why the soup base dashi had such a distinct flavor given it was usually just kombu seaweed and a little fermented tuna floating in water. He broke down the solids of the broth made from the seaweed *Laminaria japonica*, tasting as he went, until he was left with a crystalized amino acid called *glutamic acid*. He named the new flavor umami, from *umai* (うまい) "delicious" and *mi* (味) "taste." Kikunae then turned his research into a business, and his MSG-based seasoning is found on tables around the Pacific Rim. For his hard work, he is listed as one of Japan's ten greatest inventors. Not a bad outcome for a rainy-day experiment inspired by a burning desire to discover what was in his soup.

Amino acids like the one discovered by Kikunae Ikeda, and their ability to trigger the umami taste buds, are found in many traditional foods, including fermented proteins that are extremely rich in complex flavors. Cheese is blessed with abundant umami-rich compounds. Cheesemaking is ancient, depicted on the walls of Egyptian tombs and captured in the writings of Greek and Roman naturalists. Cheese is a wonderful combination of deliciousness and practicality. Cheese is milk laced with umami and frozen in time.

---

2   In particular umami is triggered by a mixture of glutamates and aspartates.

Terroir is an elusive concept that lies at the heart of why we cling to centuries-old culinary practices rather than simply munching a nutrient paste a few times a day. Terroir is the set of all environmental factors, both human and natural, that affect an agricultural product. This includes landscape, farming practice, and plant and animal growth habits. It is sometimes referred to as "the taste of the soil," but the truth of terroir is deeper than that. The term is most famously connected to wine and is that ineffable sense that grapes produced in different regions really do taste different. In fact, terroir applies whenever a crop or product has strong local variation. In addition to wine, terroir has been used to study coffee, tobacco, chocolate, agave, tomatoes, maple syrup, tea, cannabis, and — of course — cheese. If we are truly to please the Martian palate, we need to be able to bring terroir off the Earth and into space. Terroir is complexity, terroir is diversity, terroir is at the heart of the culinary language that makes up our food system.

Consider Roquefort. It is a sheep's milk cheese made in a picturesque village in the south of France. The key to making this cheese is the addition of a bacteria, *Penicillium roqueforti*, which is found in the caves under the town.[3] Only cheese aged on oak shelves within these caves may carry the name, an identity protected by European law and discerning palates. This cave aging, known as affinage, transforms the cheese. Flavor and odor compounds develop there, including acids, amino acids, fatty acids, alcohols, aldehydes, ketones, esters, lactones, furans, and terpenes. The final ones — the terpenes — reflect what the animals ate before they were milked. This witch's brew is pretty much impossible to reproduce and is so embedded in local conditions as to defy attempts by food scientists to mimic it. The resulting cheese is white, tangy, pasty, and moist,

---

3   As a rare nod to modernity, the mold is now cultured in a laboratory and added into the fresh wheels of cheese.

and is crisscrossed with blue mold. It has been made in the region for about one thousand years and comes from the Lacaune breed of sheep that has evolved alongside the cheesemakers. The sheep graze the hills and valleys on 2,100 individual farms, and their milk is used to produce about 20,000 tons of cheese each year. This cheese, by the way, has a high content of free glutamate. And so, in the end, this cheese is both the very definition of terroir but also is quintessentially umami, a fermented flavor bomb of complicated fat that owes its unique characteristics to the place that makes it.

As Lenore and Evan waxed on about the great cheese of the world, Evan paused.

"This is bothering me. Talking about umami and terroir reminds me how complex food is. What hope does the Martian farm have of ever producing great food?

"After all, Martian farmers will be starting with a sterile planet and engineered biomes bereft of the benefits of soil bacteria, centuries of culture, and careful breeding. Can we imagine anything that might actually taste good being produced there?"

Lenore gazed out the window and thought about the cheese caves she'd visited over the years. She shrugged. "It'll be tricky, but I think with a solid understanding of soil microbes, and a great genomics lab, Martian farmers will probably — over time — bring umami and terroir to the Red Planet . . ."

The challenge on Mars is like the challenge faced by the would-be maker of plant-derived cheeses back here on Earth — is it possible to coax thousands of years of complexity out of simple ingredients and then reverse engineer the delicate web of chemicals that leaves a diner yearning for more? Given what we now know about milk at the molecular level, and how it turns into cheese, could we reproduce something as complex as a blue?

Part of the answer lies in the ingredients, including the different kinds of proteins (whey and cascin mostly) that are broken down by the chymosin in rennet to form curds. It is the altered casein that gives cheese its textures and properties.[4] If we believe the ancient Greeks, one of history's first acts of chemistry produced cheese. But while the magic of cheese is partly due to the proteins, some of cheese's greatest characteristics are due to the cheesemaker's skill in manipulating the stuff. Mozzarella is made through stretching and folding. Cheddar is made by cheddaring, a fancy term for applying great pressure. Parmigiano-Reggiano's amazing flavor is partly due to the terpenes found in the mix of a morning and evening milking, but it also develops due to a brine bath and a year of affinage.

Some of these elements could be reproduced on the Red Planet without the need for a menagerie of farm animals. Although re-creating a nice Wensleydale may be beyond the ability of a Martian farmer, it is possible that if a Red Planet cheesemaker can get access to a decent supply of protein, without needing the cow on pasture, then they should be able to develop something new, something interesting, something literally out of this world.

ONE ORDER OF MILK, HOLD THE COW!

The question of producing decent cheese on Mars can be made simple: Can we take our cumulative understanding of cheese science, umami, and terroir and combine them together to produce good cheese there? After all, despite their tasty versatility, sometimes oats aren't enough. We want mozzarella that melts and stretches.

---

4   Legend suggests two origin stories. Once — a long time ago — there was a rider who stored her (or his) milk in a bag made of a sheep's stomach and was surprised to find — when they stopped for a snack — a collection of curds and whey. Another myth (and one that both Lenore and Evan prefer) is that Apollo's son Aristaio invented cheese during a bout of divine tinkering.

We want brie with bite. And when we reach for our Stilton, we want it blue. But without cows or millennia-old microbiomes, is that possible?

Answering this means we need to dig into the space-age stuff, and this requires us to shift focus and meet the newest kid on the food block: cellular agriculture (cell-ag). This emerging technology proposes to produce meat and milk but *without any animals*. Cell-ag refers to a set of technologies that fall into two rough categories. The first is tissue farming, in which meat and other products are grown in an oversized petri dish called a bioreactor.[5] The second approach is more like brewing and involves using yeast (or fungus or bacteria) that is modified to convert sugars into animal protein (instead of alcohol or other more common products of fermentation).

As we are writing, cell agriculture has exploded into the mainstream and every other day seems to bring new headlines that "yet another cell-ag start-up" had struck venture capital gold, raising hundreds of millions of dollars to bring chickenless chicken burgers, finless fish sticks, and cow-free milkshakes to the market.[6]

With all this hype, one might assume that cellular agriculture has burst from nowhere, but the idea has been slowly brewing in bioreactors for some time. New Harvest, an NGO dedicated to the development of cellular agriculture, was founded in 2004. In 2013, Professor Mark Post of Maastricht University in the Netherlands trialed the world's first cultured meat burger, and by 2014 the synthetic biology accelerator IndieBio had managed to launch several cellular agriculture start-ups. The first academic conference for cellular agriculture was hosted in July of 2016 at the Presidio in San

---

5   We will return to these technologies in the next chapter.

6   Even petfood is getting a rethink. Toronto-based Because Animals is busy culturing mouse meat for cats and rabbit meat for dogs. This is significant as pets in America consume up to 25 percent of the country's total calories from meat.

Francisco.[7] And the company Perfect Day began selling cell-ag ice cream in the U.S. during the summer of 2020, followed by cream cheese in 2021.

But the story of cell-ag reaches further back even than this. Fermentation using genetically enhanced yeasts and bacteria owes its origins to the gene-splicing techniques of Stanley Cohen and Herbert Boyer back in 1973. Cohen and Boyer's work was initially groundbreaking for medicine, as it led directly to our ability to synthesize insulin using genetically altered bacteria. Synthetic insulin became commercially available in 1982 and causes fewer allergies, has a more consistent quality, and provides a more dependable supply for diabetics, thus saving millions of lives in the past few decades alone. Synthetic insulin was so successful that it launched the entire field of biopharmaceuticals.

But let's get back to Martian dairy.

It seems almost certain that the future Martian milkshake won't rely on cows, or even oats, but rather on yeast and bacteria to synthesize the kinds of proteins we normally find in milk. These proteins will then be added to plant fats, water, and trace elements to create the look and mouthfeel of animal-derived dairy. And this technology is already being used in North America today.

For instance, when Perfect Day began selling cell-ag ice cream in 2020, they used a strain of yeast that digests particular sugars and produces milk proteins. In this way, Perfect Day uses a process more like brewing beer than milking bovines. This yeast-derived whey protein is then used as an ingredient in ice cream. Simply mix with water, fats, sweetener, and flavor and, voila: the perfect ingredients for the would-be Martian ice cream parlor where the nearest cow is 114 million kilometers away. Other companies are

---

7   Lenore sat in the back whispering "wow" and slamming back complimentary Soylent.

working on casein protein synthesis and the creation of animal fat through advanced fermentation. Our Martians might have to assemble their cheese like a jigsaw puzzle, but it is likely they will have all the needed pieces.

These technologies will be useful on both Mars and Earth where they are already solving a very cheesy problem. For the past three decades, cheesemakers everywhere have faced a critical shortage of the enzymes found in rennet, which come from the stomach lining of ruminant mammals. This is what makes milk separate into curds and whey, thus helping dairy on its journey from udder to cheeseboard.

As the global demand for cheese has grown, the supply of stomach lining couldn't keep up, so, years ago, some smart food scientists asked: could a supply of rennet be engineered? Thankfully the answer is yes. Rennet can be produced through a fermentation process very similar to the one Perfect Day uses to make their dairy proteins by employing bacteria, fungi, or yeast to convert sugars into rennet. Synthesized rennet has meant that the cheesemaking industry has been able to match demand. When fermentation-derived rennet was approved by the FDA on March 24, 1990, it was the first genetically engineered product approved for food production.

And the economics work out too: synthetic rennet has become much cheaper than the animal-derived product, and the cheese it creates has been consumed by an entire generation of happy human cheese lovers without any health or environmental problems.

This successful track record of using biotechnology to convince yeast to do the work of animals set the scene for numerous other such products including Perfect Day's products that received approval from the FDA on April 15, 2020. Their ice cream started to be sold across America as a demonstration of the technology that summer.

SAVING EARTH WITH MARTIAN TECHNOLOGY

Both of us were excited about cheese without cows, and not just for application on Mars.

"The environmental potential of these technologies is huge," Evan remarked as he paced his room.

"Stand still, will you?" asked Lenore, "you are popping in and out of frame. But yes, I agree. Dairy has such a big environmental footprint, especially large-scale dairy."

This was true. Dairy cows are a poster child for the old saying "too much of a good thing." Worldwide, milk production was estimated by the United Nations' Food and Agriculture Organization to be at 843 million tons in 2018, and this is predicted to double by 2050. Unless we radically rethink the industry, the environmental impact of dairy will double as well. And for those of us who are not planning on competing for a spot on one of the first missions to a Mars community, doubling the size and impact of our dairy industry could spell Bad News for the ecosystems we all depend on.

This is because there is ample evidence and data that show in detail how damaging grazing animals can be, highlighting their massive contributions to climate change, air pollution, damage to land, soil, and water, and the reduction of biodiversity. The major United Nations' report *Livestock's Long Shadow* (published in 2006) ranked livestock in the top two or three contributors to these problems, and highlighted that livestock is by far the largest user of the Earth's land area, equal to 26 percent of the ice-free area of the planet's landmasses. And of the crops we grow, roughly one-third are used to feed livestock.

Livestock's Long Shadow's key conclusions have been validated elsewhere including in the 2018 *EAT-Lancet Report*, published in the top medical journal of the same name. And while there are many situations where raising animals for agriculture can really help the environment, at a global level, the impact of all our cows, pigs, and chickens is expanding rapidly.

The climate impact of livestock is grim. The industry is responsible for 18 percent of all greenhouse gas emissions.[8] This is partly because there are so many grazing animals now; the biomass taken up by livestock, which is dominated by cows and pigs, is far greater than that of all other wild animals.[9] That's a lot of sirloin.

So maybe even if Perfect Day, and companies like it, never bring their product to Mars, they may help meet our rising demand for cheese and milk in a way that lowers the total environmental impact of the industry here on planet Earth. The company has tallied up the entire impact of their proteins along its life cycle and estimate that fermentation-derived dairy reduces water use by 98 percent, land use by 77 to 91 percent, energy use by 25 to 48 percent, and greenhouse gas emissions by 35 to 65 percent compared with global averages for the regular dairy sector. Given the impact of cows, this is a big deal. And given the feedstock for this process is plant sugar, yeast-derived dairy could even — at some point in the future — absorb food waste from other industries.

---

8   Lenore points out that this means if you are vegan, you can probably fly more often without guilt. You covered off your climate budget by skipping the carving station at the buffet.

9   The numbers on this are astounding. Today, there are 0.06 gigatons of humans, 0.1 gigatons of livestock, and 0.007 gigatons of wildlife. Bar-On, Y.M., Phillips, R., & Milo, R. (2018). The biomass distribution on Earth. *Proceedings of the National Academy of Sciences*, 115(25), 6506-6511. https://doi.org/10.1073/pnas.1711842115

NEW FRONTIERS

Even given these clear advantages, rolling out a new approach to terrestrial dairy will take time. But neither Evan nor Lenore expect it will be long before oat, cell, and other alternative products are readily embraced by consumers. For example, as Evan sits on a quiet Sunday morning editing this paragraph, he is drinking a coffee with oat milk, a choice he made based on flavor and mouthfeel alone, and his eighteen-year-old son prefers oat milk. Of course, in the early days, the quality of non-livestock dairy alternatives was so poor that they didn't really pose a strong threat to the industry and, until recently, the most common reason given for not switching to a plant-derived alternative was the inferior taste. But products such as Oatly are beginning to change this.

As well, in parts of the world, farmers have moved to address accusations of animal cruelty and are making sincere efforts to apply cutting-edge science to change management practices. If there is one thing that's true about agriculture, it's that there are environmentally good and bad ways of producing pretty much everything, and there are a lot of dairy farmers all over the world who are working hard to keep their animals well-cared-for and act as careful stewards of the land.

And since the dairy industry enjoys the fact that milk and cheese are associated in the minds of many consumers with good health, and that it is subsidized by the government in most major Western countries, it is understandably defensive of the shift to plant-derived and yeast-derived dairy. The lawsuit against Oatly was just an opening volley. The dairy industry is all too aware that plant-derived products threaten to take market share from them. They know, for example, that there has been a decline in the amount of fluid milk North Americans consume. So, the industry has gone to the courts. In the EU, the term milk cannot be used for drinks that are not made from mammary secretions (though coconut milk and almond milk are, oddly enough, exempt). In the United States, the National Milk Producers began to lobby extensively in the early 2000s to ensure that the Food and Drug Administration keeps defining milk as "the lacteal secretion, practically free from colostrum, obtained by the complete milking of one or more healthy cows."[10] There is almost certainly more to come.

Regardless of the court cases, or marketing campaigns designed to win over the public's allegiance, one thing is certain. Between the rise of plant-based alternatives and the rapid maturation of the technologies behind cellular agriculture, it seems we will very soon have dairy substitutes that check all the boxes for our dinner on Mars. In short, we can expect Martian communities will have extensive cellular dairies, churning out butter and ice cream and even cheese.

Could such a cheese ever go head-to-head with cave-aged Roquefort? Lenore and Evan slightly disagree on this point. Lenore is bullish and thinks that this will happen with time. Evan is a bit more traditionalist. While he foresees these alternatives gaining a bigger share in the market on Earth — and really being one of the bases of anything to do with space exploration — he can't see a lab reproducing the full range of tastes that come from 10,000 years of experimentation.

But maybe Lenore's optimism is better founded? Yeast-derived dairy is basically a ferment, and there are certainly ferments on Earth that are dripping with umami and terroir. These include wine, beer, and distilled products like Scotch whiskey. At top-end restaurants such as Momofuku and Noma, fermentation teams even talk about microbial terroir. So maybe — just maybe — each Martian

---

10 https://www.accessdata.fda.gov/scripts/cdrh/cfdocs/cfcfr/cfrsearch.cfm?fr=131.110

community might, in the distant future, have its own populations of yeast and bacteria, much as San Francisco boasts its own species of yeast, *Lactobacillus sanfranciscensis?*[11] If we take along our microbial companion species, the future for umami and terroir on Mars might yet be bright and tasty.

---

11 They owe their distinctive sourdough bread to this local species.

CHAPTER 7:

The Fish of the Sea and the Birds of the Heavens

TO CATCH (OR PRINT) A FISH

In some ways, Lenore's life can be mapped through her relationship with meat. She's a fisherman's daughter, and in her earlier years, she thought all reasonable adults owned boats of various sorts. In college, she worked a variety of non-glamorous food-industry jobs. As she learned more about how animals in the food system are treated (in part thanks to her well-informed vegan partner), she shifted her own diet towards the plant side of the spectrum. As her career took her into massive factory farms and slaughterhouses, her resolve deepened.

Evan is a bit more on the fence. While he basically agrees with Lenore, his position is that animals play a vital role in recycling nutrients in well-managed farming systems, and raising animals is a source of livelihood for hundreds of millions of people (including many of the world's poor). He also respects the fact that consuming animals is intrinsic to many cultures, including Indigenous communities. Yet he completely agrees with Lenore that many of the

ways animals are raised and slaughtered today are deeply problematic from both animal welfare and sustainability perspectives.[1] So,
while he still enjoys eating meat occasionally, he's pretty discrimi-
nating about where it comes from.

With that said, the two of us have had a running disagreement
around the role of meat in space. This tension emerged early in
writing this book. Lenore opened the debate.

"Evan, I'm wondering if we should even bother to look at meat?
Advanced fermentation of dairy is one thing, but couldn't our
Martians simply be vegetarians?" As she said this, she was staring
out the window of her apartment in Vancouver where the crocuses
were coming up and the dark of 2021 seemed to be receding.

Meanwhile, in Ontario, winter was providing another unwel-
come snowfall. "Humor me; let's stick with the protein theme for
the moment," said Evan, glancing out at the accumulating inches
of icy white. "Are Martians going to eat steak produced locally?
Seared tuna, wild-caught on Earth, and shipped up a space elevator
in vacuum packs? I mean, we can't just have our community living
on vats of algae."

"But why would we take animal farming, which is horridly inefficient and ethically bankrupt, with us to a new planet? Let's try to
figure out not only how BaseTown is going to get its protein but
look at ways that can run on minimal inputs and still taste great."

"Fair," Evan said. "I'm all over being 'plant forward' in terms of
diet, but I don't know if we'll be able to sell Mars if it's vegan. Most
of the world is omnivorous . . . convincing would-be Martians to

---

1   As noted elsewhere in this book, Evan's first major non-academic writing proj-
    ect was a book entitled *Beef: The Untold Story of How Milk, Meat, and Muscle
    Shaped the World* (William Morrow, 2008). It's an unapologetic celebration of
    steak and ice cream but also explores the very serious sustainability issues linked
    with animal agriculture.

sign up for a months-long journey through the inky void will be hard enough, so, I think we need to offer food that most people would consider to be halfway decent."

Lenore paused. "Well, if we are making dairy without cows, maybe we just take that a step further. I know a guy from Seattle who may be able to help. And I do miss salmon . . ."

Lenore and her partner, Katya, often make a day trip or weekend wander to Seattle. As a geographer and incorrigible traveler, whenever she passes the customs offices that separate the U.S. and Canada, Lenore marvels at the strange power of the imaginary line separating the two countries. After all, Seattle and Vancouver are both washed by the waters of the Salish Sea. The Salish peoples, who occupied the region for thousands of years, had their own boundaries and borders, and the sea itself was a combination highway and larder. The twisted topography of the Pacific Northwest doesn't lend itself to easy habitation, but the sea was a gift of open horizon, smooth passage, and oceanic bounty. Lenore's family sold fish in Seattle, tracing these ancient trade routes and traversing the modernized portage of shipping locks to Lake Union. Even all these millennia later, the sea is still a highway.

Lenore and Katya's trips to Seattle follow a pattern. They stop at Trader Joe's to stock up on exotic foods (cookie butter and cauliflower gnocchi, anyone?).[2] They poke around the shops on Capital Hill. And Lenore visits Pike Place Market, where she gathers truffle salt, hand-ground spices unknown to the spice-averse palates of many Canadians, and cocktail bitters. And she stops, for a moment,

---

2   People sometimes ask Lenore what she misses about living in San Francisco. Though fog, the smell of Eucalyptus, Victorian houses on the hill, and the Golden Gate top the list, Trader Joe's probably makes the top ten.

to watch the vendors throw salmon. It's something to behold. Brawny, bearded men bent double over massive, gutted fish that are displayed carefully on mounds of ice. They grasp the slippery salmon, and with a confident underhanded toss, throw the bodies twenty feet, over aisles, counters, and other consumers so they are caught by equally burly fishmongers who weigh and bag the fish. The assembled crowd cheers at the sight.

The short explanation for one of this strangest of tourist rituals is that decades ago, fish stall owner John Yokoyama noticed that, thanks to the market's vintage architecture, he had to take one hundred steps to move a fish from the back of his shop to the front counter. Noting the distance was shorter as the crow flies, he tossed a fish through the air at his assistant. The technique caught on, and now people from around the world come to watch.

The longer explanation for Pike Market's fish tossing tells us something about food that we will need to know as we prepare for Mars. Seattle is a water place, a boundary place. Here, at the heart of the Salish Sea, people know and love the salmon. Salmon are a keystone species, cherished by tourists and locals alike. Yet despite this love (or because of it) the salmon are in decline — a combination of habitat loss, decades of overfishing, pollution, and climate change is pushing this species into danger. And so, a group of innovators is working to save the salmon by developing new technologies that will allow us to satisfy our craving for salmon without depleting the ocean. From this, we can glimpse at how we might bring protein to Mars.

TEACH A PERSON TO FISH AND YOU FEED THEM TILL THE STOCKS COLLAPSE . . .

There is something magical about seafood. It is one of the few sectors of the food economy still connecting us to the hunter-gatherer lifestyle that dominated most of human history, where people actively consume food harvested wild. And it isn't a marginal activity; today, fish accounts for 17 percent of all animal protein devoured by humans, and about half of this still comes from the wild. Of course, this is changing, and farmed fishing is growing by leaps and bounds, but either wild or farmed, all this fish is a big business. Nearly sixty million people work in fisheries and aquaculture, 200 million jobs are directly or indirectly connected with harvesting fish, and about half of this trade comes from developing countries.

All this activity takes a toll on wild stocks. Roughly two-thirds of the world's fisheries are either fished to their limit or over-fished. But this isn't entirely a bad news story. A recent study led by University of Washington professor Ray Hilborn showed that in regions where fisheries are carefully managed, stocks are stable.[3] In some cases, well-managed fisheries are improving. Careful, sustainable management has spread to over half of the Earth's fisheries. Fish species aren't out of the aquatic woods yet, but we've begun a transition from raw exploitation to stewarding complex ecological systems. These are the skills we will need to perfect if we are to survive in the much smaller ecosystems we will establish on Mars.

The salmon faces some particularly difficult challenges. There are five commercial species of salmon swimming in the waters of the Pacific Northwest: chinook (also known as king), coho, chum, pink, and sockeye. The first and last of these are most highly prized for their culinary value, but each of the species makes a fine meal. The sockeye is known for its bright red flesh and delicate taste. Overfishing has been a problem for the salmon of the West Coast

---

3   Hilborn, R., Amoroso, R.O., Anderson, C.M., Baum, J.K., Branch, T.A., Costello, C., ... & Ye, Y. (2020). Effective fisheries management instrumental in improving fish stock status. *Proceedings of the National Academy of Sciences*, 117(4), 2218-2224. https://www.pnas.org/content/117/4/2218

at least since Europeans arrived. They command high prices at the dock and demand is always strong.

Salmon face another problem. They are anadromous, meaning the fish are born in freshwater, migrate to the ocean to live their lives, and then return to their home rivers to spawn and die four years later. Near the locks leading to Lake Union in Seattle, the Ballard Fish Ladder provides a shortcut for the salmon allowing them a quick route out from the Pacific and into their freshwater spawning grounds. Watching the fish leap up the ladder, moving from pool to pool, is almost as popular a fall activity for citizens of the Emerald City as watching the fishmongers in Pike Market toss their less fortunate cousins. The settlements of North America's West Coast even today are often located on major salmon rivers, reflecting the historical convenience of an abundant food source that comes to the diner, like clockwork, every year.

But the salmon's migratory pattern makes them particularly vulnerable to human activity. In the ocean, they face the same over-fishing, pollution, and climate challenges as other fish species, but because they need the inland area to reproduce, they also require healthy rivers full of quiet wetlands, meandering tributaries, undisturbed gravel, and cool clean water. Human habitation doesn't tend to mix well with pristine rivers. From housing subdivisions disturbing the quiet swamps and meanders of local streams, to the hydroelectric dams on the region's great rivers, we have altered the habitat West Coast salmon depend upon. Many of the once teeming salmon runs are now at about 5 percent of their historical maximums.

Climate change is making the situation worse, as river flows are more unpredictable with more common floods and droughts. As a result, about half of the salmon consumed on the West Coast is now farmed in giant net pens that are suspended in the ocean. But these too come at a cost since fish farmers generally raise the more resilient and faster-growing Atlantic salmon, which makes financial sense to the fish farmer but also means that the diseases common to Atlantic salmon (such as sea lice) now spread to their west coast cousins in the wild.

It is hard to imagine the west coast of North America without its iconic fish. Salmon were harvested intensively by Indigenous groups of the Salish Sea and were so important in part because of their anadromous nature. Their convenient habit of returning to their home streams meant that Indigenous communities in the region built elaborate settlements near rivers and let the abundant fish come to them. The peoples of the Pacific Northwest stored this harvest every fall, smoking fish, moving fish inland to dry it in the hot river canyons beyond the coastal mountains, and burying the fish to ferment in pits. They pounded dried salmon and mixed it with grease and dried berries to make pemmican, which was traded far inland. Today, we are just as creative, and salmon is roasted, cooked on planks, smoked, candied (a popular purchase at the airport for departing tourists), served in burgers, and used for sushi.[4] The BC roll is a Vancouver staple invented by local master chef Hidekazu Tojo, who used salmon skin to replace the difficult-to-source eel meat found in more traditional preparations. Salmon is so popular that it remains the most valuable fishery in Canada, and on the U.S. side of the border, the fishery generates about 1.2 billion dollars in GDP and supports 17,000 jobs.

"Yes, but . . ." Evan interrupted Lenore's trip down Salmon Lane, "what does all this have to do with a book about food on Mars?"

Lenore replied, "If this tasty bounty is to continue, the fishery must be stabilized at a sustainable level. But where, then, can we source more salmon? After all, demand is rising everywhere.

---

4   Salmon sushi was never served in Japan, oddly. It was first developed in Norway, and many of the key recipes were perfected in the Pacific Northwest. It then migrated back to Japan in the 1990s.

"I think that new technologies may mean that it should be possible to enjoy salmon, only without the impact on the fish themselves. Just before the pandemic, I talked with Justin Kolbeck and Aryé Elfenbein who founded a company called Wildtype. They use cellular agriculture to create salmon by farming fish cells. Unfortunately, the stupid pandemic means I've not yet been able to taste their product, but if you check out their web page, the product looks perfect!"

Evan pulled up their web page and nodded. The fish looked mouth-wateringly good, just the way sashimi should.

"Is that a trick?" he asked, "I mean that looks just like the real thing."

"Apparently it's legit. And this is all cell-grown protein and fat. It's supposed to taste great. And is complete without microplastics, mercury, or, of course, fish slaughter. Wildtype is even opening a tasting room in San Francisco."

And the two of us slid into daydreaming about traveling down to taste the final product.

. . . BUT TEACH A PERSON TO PRINT A FISH AND YOU JUST MIGHT SAVE THE OCEANS

The point of all this discussion is that either through cell cultures such as the sort Wildtype uses, or the dairy fermentation described in the previous chapter, cellular agriculture offers the possibility of radically reducing our reliance on captive and wild animal populations. While this could seriously upset the fishmongers of Pike Place Market — and fishers, ranchers, shepherds, and feedlot owners around the world — the possible benefits of this technology could be a better-managed environment, along with life-changing improvements in the ethical treatment of animals. And these technologies could really support a community in space.

The process of creating meat using Wildtype's techniques is surprisingly easy to describe, though not at all easy to do. First, one selects an animal and does a biopsy to gather a small sample of cells. One then sorts out the stem cells, and, through a process of cell differentiation,[5] the desired cells are activated to create the tissue cells. The results are placed inside a bioreactor (imagine a big vat filled with a nutrient broth) and allowed to divide. To speed the whole process up, one can add growth factors, such as the proteins produced in the fancy barley grown in the Icelandic greenhouse described earlier.

This results in a mixture that is basically made up of the targeted tissue cells. Although the basic ingredients are there, this slurry lacks texture, mouthfeel, taste, and form. So, the next step is tricky. Because meat is three-dimensional, a structure must be imposed. Some companies are attempting to do this by building lattices or scaffolds that guide cell growth. Other companies just create a thick goopy semi-congealed liquid and use 3D printers to extrude a final form. Either way, this is much less simple than it sounds, and getting the broth of protein molecules to form into something a meat-eater would consider a suitable replacement for a piece of salmon sashimi is currently limiting companies like Wildtype from getting to market. However Wildtype is well on the way and will soon be providing their product to the consumer; and when they do solve this problem, we will have taken a major step to finding a product that has every chance of reducing pressures on wild populations here on Earth. And we will have created the possibility that future Martians will be able to enjoy sashimi.

But before we get too confident cell-ag is about to replace animals, there are other challenges that scientists must overcome. The

---

5   This process alone is incredibly complex. The discovery was made in the 1870s by Walther Flemming, and we are still unraveling the mysteries of how cells develop.

cells used to create the animal proteins naturally divide only fifty times. So, unless one creates an immortal cell line, or repeatedly biopsies live animals to gather new cells, there is only so far you can take this process.

As well, the serum used to feed the cells as they grow in the bioreactor is both expensive and difficult to create; it is a mix of amino acids, sugars, lipids, and hormones. At first, fetal bovine serum was used, though grinding up cow fetuses isn't the most animal-friendly method, and it means the stuff isn't vegetarian. Vegan serums are now being developed, and they are needed at scale. But these are not insurmountable problems. For instance, companies such as Eat Just and Future Fields are working on churning out bulk plant-based serum and the whole industry is — at the time of writing — starting to get big enough to be interesting. Bioreactors are easier to manufacture now, as they are basically brewing vats for meat. There are vats in Israel that can produce a volume equivalent to what 1,500 chickens would produce in just a few weeks. And chemists and brewers already know how to keep the environment within these biofoundries stable. And all these steps, resting on a century of biological science, must be scaled up. This is coming. Eat Just started selling chicken nuggets that contain cellular chicken in Singapore. North American products aren't far behind.

These innovations haven't emerged overnight. Growing protein without animals began to leave the realm of fiction in the early twentieth century when Yale zoologist Ross Granville Garrison successfully cultivated living tissue. That work rested on earlier advances as well. In 1885, Wilhelm Roux kept chicken cells alive for several days, making him the world's first cellular farmer.

Despite these early breakthroughs, cellular production of protein remained a novelty for decades. The first documented tasting of such a product occurred in March of 2003 in Nantes, France, when Australian-based artists Oron Catts and Ionat Zurr created a cutlet of cultured frog meat in an exposition called Disembodied Cuisine. However, this was mostly an art project, so we need to turn to the life's work of the Dutch doctor Willem van Eelen, one of the key pioneers in the field.

Born in Dutch-controlled Indonesia in 1923, van Eelen was interned in a POW camp by the Japanese during World War Two, where he almost died of malnutrition. Van Eelen was struck by the cruelty his captors showed to both human and non-human residents. He took from these experiences a deep interest in food and a moral responsibility to end human and animal suffering. Following the liberation of the camps by the Allies, van Eelen returned to the Netherlands and studied psychology and medicine. Decades later, in the 1990s, he entered a partnership to create an in vitro meat process, and though the laboratory work was not as successful as he had hoped, van Eelen filed several patents. In 2000, he formed a consortium of Dutch researchers and secured funding for four intense years of research between 2005 and 2009 that set the stage for the modern renaissance in cellular agriculture.

This is where vascular physiology professor Mark Post at Maastricht University steps into the story and the tale of the world's biggest celebrity burger begins. Post realized that the cool science wasn't enough and that the funders needed an ambitious vision everyone could get excited about — a live presentation of actual cellular meat. After briefly considering growing a sausage, and at the urging of his funder Google co-founder Sergey Brin, he settled on that most American of meals, the hamburger. Post's first cellular burger cost a quarter-million Euros to produce.[6] The

---

6   In reality two burgers were produced, the one destined for Prime Time and a
     spare. That spare was plasticized and is displayed at the Boerhaave Museum
     in Leiden.

process was arduous; over three months, 20,000 individual muscle fibers were grown from bovine stem cells collected from the flesh of a Belgian Blue cow, a breed that has a mutation allowing it to produce so much muscle that the calves almost always need to be birthed via cesarian section. The cells harvested from the Belgian Blues were grown in a bath of bovine calf serum and then extracted one at a time. Finally, they were pressed together to form the burgers. The resulting meat was completely lean and whitish/gray in color. Post mixed in saffron and beet juice to make the finished product a little more photogenic.

On August 5, 2013, Post's team brought the world's first test-tube burger to a television studio in London for the moment of truth. In front of the press, chef Richard McGeown of Couch's Great House Restaurant in the fishing village of Polperro, Cornwall, did the honors. He prepared the burger in a little bit of sunflower oil and butter. The finished product was then plated with no garnish and tasted by futurefoodstudio's Hanni Rützler, who described the burger as dry but indistinguishable from meat. Next on the judging panel was author Josh Schonwald, who was impressed with mouthfeel and texture, comparing the cell patty to a typical fast food hamburger, only without the fat. Post himself then sampled his lab's example of very, very slow food.

Publicity stunts have their place. When the lights went down and the dishes were tidied, the world's attention was firmly on the cellular production of meat. Mark Post went on to found Mosa Meat and by 2020 was boasting he could produce the same burger patty for about ten Euros. Mosa is currently working on perfecting producing cellular animal fat (to mix in with the muscle cells) and is also working to perfect the process of brewing vegan growth serums. Post admits to being on a crusade to ensure that there are a lot fewer cows on the planet in the future. He argues that today's cattle population of one and a half billion cows needs to drop to about 30,000.

"Hang on, are you suggesting that if cellular agriculture becomes the standard on Mars, we might see a massive drawdown in animal agriculture on Earth?" Evan looked concerned, as if his steak were about to vanish into the mists of history. "I mean I know the industry needs to change, and we all must eat less meat, but by that much? Don't forget, my first book was called *Beef: The Untold Story of How Milk, Meat, and Muscle Shaped the World*, and I know the cultural place that beef has for people all over the world."

"I hope so," came Lenore's quick reply. "Animal agriculture uses huge amounts of land, energy, and water. I know you think there are some situations where animals can be raised in ways that meet environmental or ethical standards but, overall, the way the world produces animals today is madness. If I'm going to eat fish on Mars, I want it to be grown in the lab and printed by a 3D printer."

If Lenore is dreaming of eating omakase in a Martian crater, several companies are working to make this dream a reality here. These include Finless Foods, which is working on replicating bluefin tuna; BlueNalu, which is planning to offer a variety of seafoods; Shiok Meats, a company pioneering the production of shrimp, crab, and lobster; and, of course, Wildtype, among others. Finless even sent cells to the International Space Station where they were cultured and shaped into spheres using a 3D printer. The oceans of Mars might be in a lab. On Mars, we predict that the local sushi bar will plate cellular fish together with small amounts of locally produced rice and seaweed that comes from large bioreactors designed to produce algae.

But ultimately, fish isn't the world's most popular protein. That honor falls upon the humble chicken.

WINNER, WINNER, CHICKEN DINNER

One of Lenore's favorite creatures, the humble chicken, was first domesticated from a wild red junglefowl in Southeast Asia some eight thousand years ago.[7] This dooryard bird was kept mostly for eggs until the early years of the twentieth century, but today, chicken dominates the animal protein category. At any given time, there are twenty-three billion chickens pecking away somewhere on the planet, destined to serve our insatiable desire for cheap protein.

We owe these birds a deep apology. Few species have suffered so much through domestication, and that suffering grows every year as demand grows. As geographer Carys Bennett[8] and her team note, the modern broiler chicken embodies humanity's giant impact on the biosphere. The chicken has been a food system workhorse for a long time, first becoming a popular culinary addition after the Egyptians learned how to incubate eggs, freeing hens to lay more. Attendants kept the eggs warm by burning small fires of straw, carefully adjusting the heat. And in truth, modern methods are not that much different.

On the culinary side, the Romans developed many recipes involving the birds that would seem quite acceptable on tables today; they invented both the omelet and the tradition of roasting

---

7 The reality isn't quite that simple, as several other wild birds have been bred into the domestic chicken over the millennia. But *Gallus gallus* and the domestic chicken are pretty close family members.

8 Her article "The broiler chicken as a signal of a human reconfigured biosphere" is available at https://royalsocietypublishing.org/doi/10.1098/rsos.180325

stuffed carcasses. Under Roman care, chickens got bigger (proba bly due to a combination of breeding as well as better feeding), but when the empire collapsed, so too did the first era of the chicken. It wouldn't be until *Gallus gallus domesticus* reached America that they began their next period of culinary ascendency. From the kitchen gardens of Long Island, this bird is at the center of one of the most mechanized, efficient, and, in many cases, cruel animal industries in all of history.

And what a rise. In 2016, their standing population makes them the most numerous bird on Earth. There are, by contrast, only a quarter-billion pigeons. In Europe, chickens alone outnumber the combined population of all wild bird species. There are likely more chickens alive today than any other bird in the history of the planet. And these modern chickens are efficient. It takes two kilograms of feed to produce one kilogram of live chicken, a ratio that has fallen by half since the end of World War II. And they are fast: a chick matures into a full-sized broiler, ready for plucking and the stew pot, in just six weeks — about half the time it took before the results of selective breeding resulted in the behemoths that now adorn our supermarkets.

The cost of this efficiency is discomforting. Today's domesti-cated chicken is twice as big as the average chicken raised during WWII and weighs five times as much. Their bone structure is rad-ically different. A broiler cannot survive without human help; they can no longer forage successfully. If allowed to live past six weeks, the gigantic size of a broiler begins to radically decrease the bird's lifespan as its heart and lungs struggle and its muscles strain under the weight. The modern broiler chicken, once proudly prodded along by programs such as the USDA's Chicken of Tomorrow, is a colossus, bred for a short life, and raised to ensure that it puts on the maximum amount of meat with the fewest inputs. Animal welfare problems are legion.

On this, Evan agreed. "The chicken industry definitely has deep problems. Animal welfare, health, environment. Even if we could take chickens to Mars, we probably shouldn't."

Lenore nodded. "Yup, and the weirdest part is the average consumer only wants the white meat. They don't want their chicken to taste like chicken or look like chicken. No wonder cellular agriculture companies are so focused on chicken as a product."

For the sake of argument, let's say we wanted to bring chickens along to Mars. Let's imagine we were going to treat them well, letting them range in our agricultural domes in search of a carefully curated diet of greens and insects. They make a nice pet, after all. We could gather their eggs and enjoy a space omelet or two. Could these birds make the journey? Birds, it seems, don't enjoy space very much. On March 22, 1990, four adult quail were taken by Soyuz TM-10 to the Mir space station by flight engineer Genandy Strekalov. One of the females laid an egg on the journey, joining a set of thirty-two fertilized eggs brought from Earth. When the first of these eggs hatched, the healthy quail chick became the first vertebrate to be born outside the Earth's atmosphere.

But that early win aside, the chicks found zero gravity very difficult. The quails couldn't perch to feed and had to be hand-fed. The cosmonauts devised clever little harnesses for the birds, successfully building a high-altitude quail farm. The birds suffered numerous problems with muscle loss and hormonal difficulty, though one hen did survive the return to Earth, and recovered to join a terrestrial flock. So, chickens, with difficulty, could be taken to Mars, where the gravity is likely strong enough.

Perfecting chicken substitutes for the human Martians could greatly reduce the resource demand of a very popular protein on both planets. Consumption of chicken has tripled in the U.S. since 1960, and global consumption rose 30 percent in the first decade of the twenty-first century alone. In part, this is because of successful marketing; chicken is viewed as a healthy substitute for red meat. And while chickens require much less energy, water, and feed per edible kilogram than cattle or pigs, they are still much less efficient than cold-blooded salmon. And the greenhouse gas emissions of a pound of chicken are ten times higher than for the equivalent amount of grains or pulses.

For the plant-based company Eat Just, the egg came first. The company was founded in 2011 by Josh Balk and Josh Tetrick in Los Angeles. At the time, their company was called Hampton Creek, but ever since, Eat Just has followed a vigorous program of work that would make sense to the plant-hunting explorers of the Victorian era; they tested countless plant proteins, meticulously recorded their results, and explored those with similarities to chicken and eggs. They logged each plant in their database, Orchard. Their Just Mayo went to market in 2013, spurring a round of expansion and fundraising.

Fending off a legal challenge in 2014 from Unilever, who attacked them in court because they tried to market themselves as producing mayonnaise despite not using eggs, the company reached an important compromise with the FDA: they could call their product mayo but only so long as the label made their eggless status clear. And this launched Eat Just into that special category of legendary companies — unicorns — that became so profitable that early investors became very wealthy. And then, having clearly proven their ability to produce egg proteins without chickens, Eat Just took on chicken itself.

In 2017, Eat Just announced they would be growing chicken nuggets using cellular agriculture. The result was 70 percent synthetic (cellular) meat, with the remainder plant-based materials that give the product the right structure and texture. This product is destined to be as historic as the quail cosmonauts; by late 2019, the cost of each chicken nugget had fallen out of the stratosphere to only fifty dollars per nugget. Since then, the price point has continued to plummet. In December of 2020, the Government of Singapore approved Eat Just's chicken for public sale, and restaurant 1880 became the first place on Earth where one could find cultured meat on the menu. The next step on the trajectory won't be a food court on Mars, but the technology is now pretty much ready to allow future astronauts to enjoy a locally produced chicken burger without the hassle and animal welfare implications of trying to raise these birds on the Red Planet itself.

THE PLANT PARADOX

Hearing all this high-tech stuff, however, was starting to strike Evan as perhaps missing the point. He called Lenore.

"Look, I'm having second thoughts. Maybe fermentation and cell-based production could replace animal agriculture on Mars and on Earth, but plants are, after all, quite capable of making healthy protein. Maybe you were right at the outset. Maybe future Martians will just go vegan?"

Lenore smiled. "Sure, sure . . . so, now you come around to that? Well, it's possible, but I think you were right the first time. A little bit of cultured meat will provide some much-needed variety in BaseTown's refectories and eating establishments."

Our reading of the tea leaves concludes that there are at least two reasons some sort of meat or meat analogue will always play a significant role on Earth as well as in the Martian diet: one practical, one cultural. In terms of culture, most of us have deeply personal relationships with food. We care about our cherished foods and often talk about foods as being "authentic," "traditional," or "natural," all notoriously vague terms that reflect the degree to which something is roughly what we believe it should be. For many cultures, meat is deeply entrenched, paradoxically in part because meat-eating was much less frequent in the past. Because meat was hard to get, expensive and rare, it was associated with life's most important meals: feasts, holidays, weddings, and other major celebrations. Just consider the adjectives "meaty" and "vegetative" and you get a sense of meat's place in our society.[9]

Other foods are, of course, linked to place. As we've discussed earlier, cheese carries astoundingly deep cultural meanings that are attached to the breed of ruminant, specific forage patterns, weather, history, and the individual cheesemaker. Meat is no different. Even post-millennials, many of whom are likely to embrace flexitarian diets, report loyalty to products that carry good memories and that are associated with elements of identity and belonging. This effect has been called gastronationalism in which attachment to food can override practical concerns including animal welfare, social justice, and environmental impact, leading to protectionist actions that can even overpower economic considerations. France's love of foie gras is one example, with the French resisting all attempts to reform production even as the product has been banned in many jurisdictions around the world.

If cellular agriculture is to succeed, therefore, it will likely need to work hard to avoid challenging consumers' deeply held cultural values around the central role of meat. After all, people don't eat meat specifically because it once came from a living animal or even animal cells, they eat meat because of the sensory experience the product evokes and the memories the food triggers. They also eat meat out of habit, a topic that could fill a book on its own. Therefore, if good quality animal analogues hit the market (and all the evidence

---

9   Evan's first book, *Beef* – co-written with journalist Andrew Rimas – goes into these points in great detail.

is that the industry is taking breathtaking strides towards this), then consumers may be more receptive to the ecological and moral critiques of meat and dairy. And early evidence suggests that even entrenched consumer habits may be changing as plant-based diets are rapidly growing in popularity.

The more practical reason for our predictions that these animal analogue products are destined to become mainstream is because of economics and marketing. Plant-based substitutes for meat are getting a lot of press, they are getting tastier, and they are getting cheaper. While plant-based foods that act as animal analogues are not new, the next generation of these products seem to be crossing a frontier akin to the quest for artificial intelligence: meat substitutes so realistic they meet a Turing test as indistinguishable from the real thing.

Lenore was in her kitchen, deep in the pandemic ritual of prepping too much food. She was busy whipping up a batch of Impossible balls made with Impossible Food's ground round product, using, in a delightful irony, her grandmother's secret meatball recipe.

Lenore remembers her first encounter with the product. She was in Boston for a conference on food technology and set out on a side mission — to finally try Impossible Food's ground beef replacement. In Canada, Impossible Foods had been, well, impossible to find, but in Boston, the restaurant, Clover, was serving up pita sandwiches stuffed with Impossible balls in a red sauce. Lenore braved the blistering heat and the jammed trains of the Boston T to secure her first bite of the latest in food technology. And at that moment, with that preparation, Lenore encountered her first plant-based meat to pass the culinary Turing test. Despite the heat, she ordered another, taking furtive bites on the T ride back.

The story of Impossible Foods begins in 2009 when Stanford biochemist Patrick Brown turned his attention to the problem of intensive animal farming. Convinced the answer to reducing the impact of meat lay in providing delicious alternatives at a reasonable price, Brown left academia for industry and began Impossible Foods. Their company launched the first Impossible Burger in 2016. The 2.0 version arrived at lunch counters in 2019 and reduced both the sodium and saturated fat content, while preserving the environmental benefits of the original version. Though not yet at cost parity, the cost of Impossible's products continues to drop.

The reason Impossible can fool someone who enjoys food as much as Lenore is a not-so-secret ingredient called a heme replacement. Heme is what makes meat red, but it is also found in plants. Impossible uses heme found in soy, but companies are increasingly making their own heme using a process similar to the fermentation discussed in the last chapter. Heme replacement was approved for restaurant use by the FDA in July of 2018 and for home use in the fall of 2019, paving the way for delicious vegan meatball sandwiches. Much as with Just Foods, Impossible maintains a library of plant proteins and fats and continues to tinker with their offerings.

Impossible has come a long way from its first availability at Momofuku Nishi in New York. It is now widely available in multiple countries, and the company has attracted over 1.3 billion dollars in capital.

Evan hadn't had the opportunity to try Impossible Burgers until one day, after leaving the grocery store, he drove past a Burger King advertising the Impossible Whopper. Pulling into the drive-through, he put in an order, feeling a bit like a teenager buying condoms or booze. After all, he's a professor of food sustainability, and the city of Guelph isn't all that big. What if one of his students saw what he was doing? Sitting in the parking lot of Burger King, he pulled back the greasy wrapper and stuffed a couple onion rings in his mouth. He ate a bite of the burger, cleansed his palette with another onion ring, then pulled out a bit of the patty, broke it off, and closed his eyes to better concentrate on the taste. He pulled out his phone and texted Lenore:

"I've tasted the future, and it tastes . . . exactly like the past."

Patrick Brown still thinks that animal agriculture can be eliminated by 2035, allowing land reserved for animal agriculture to return to the wild. Could Impossible conquer the domed crater habitats of Mars? Maybe. But this could be a vision for the future of meat here on Earth, and sooner than we think: alternative proteins are already worth over two billion dollars a year in market share, and they continue to grow, driven by venture capital backing, customer demand, as well as advances in genomics, molecular biology, and food science. And we are nowhere near the market cap; the estimated value of the animal protein market today is 1.7 trillion dollars and growing. Alternative protein could reach 11 percent of that market by 2035, rising rapidly after that.[10] In 2020, over one and a half billion dollars in investment capital poured into emerging companies, and over fifty cellular agriculture companies are active globally producing analogues for beef (Mosa Meat, Aelph Eatery, Upside Foods), chicken (Eat Just, SuperMeat), seafood (Wildtype, Finless Foods, BlueNalu, Shiok), leather (Modern Meadow), gelatin (Geltor), dairy (Perfect Day, TurtleTree), and eggs (The EVERY Company, Eat Just). But the industry, aside from some outliers, is concentrated in a few key geographic areas, including California's Bay Area, Singapore, Israel, and the Netherland's Golden Triangle. It has yet to really go mainstream across the globe.

THE ANIMAL ANALOGUE FUTURE

Evan and Lenore recapped their exploration of Martian protein, with Evan leading off, "So perhaps the Martian table will be spread

---

10 https://www.bcg.com/en-us/publications/2021/the-benefits-of-plant-based-meats summarizes several transition scenarios

with foods that utilize plant-based technologies, fermentation, and cellular agriculture, sometimes even all together in the same bite?"

Lenore nodded. "I think so. Our relationship with meat is constantly changing, and as we've already discussed, there are animals that used to be historically important, like horses, that have vanished from our day-to-day lives. Just because everyone once rode horses doesn't mean we'll take horses to Mars. And Martians most certainly won't worry about a whale oil shortage, even though New Bedford, Massachusetts, was once known as 'the city that lights the world' because it was the center of the whaling industry. So just because we eat a lot of cows, pigs, and chickens today doesn't mean we'll eat a lot of them in the future, and I certainly don't think we'll import these creatures to a different planet. Even though a shift away from animal agriculture now looks radical and disruptive, I personally think it is also completely expected given the larger arc of history."

"I've come to see your point," Evan replied. "I really don't think there will be many, if any, animals on Mars, and that end of an epoch will likely happen on Earth too. With that said, it's not as though horses or sailing vessels have vanished from Earth entirely, but they are now mostly recreational things used by affluent folks, and not regular parts of everyday life."

Evan could see Lenore grinning.

"What?" he asked, sensing she was leading to a punch line. "Are you feeling smug that you've convinced me?"

"Yes, but there is another element to this argument that just occurred to me . . . I have a feeling that there is another creature that won't be as plentiful on Mars. And by replacing it, we'll also have a huge impact here on Earth."

Evan shrugged, "Ok, you've got me . . . Zebras? Raccoons? I give up."

"On Mars, there won't be, and can't be, many human farmers."

PART IV:

RED DAWN

CHAPTER 8:
Old MacDonald Had an iFarm

FARMER 5.0

L abor is generally a matter of economics. If people are few — or if those people have a lot of other demands on their time — then there are big incentives to invest in automation. This has been a driver for both manufacturing and farming, and machines now do much of the work laborers used to do. For the past hundred years, the number of farms and farmers has been steadily declining as those remaining on the land invest in bigger tractors, while farm kids all over the world decide there are better options in the city. It's hard to imagine Mars being any different.

Evan should know. His grandfather was a fruit and vegetable farmer not far from Niagara Falls in Ontario. One evening when Lenore and Evan had just become acquainted, Evan told her how he ended up a professor of food systems.

"I remember one afternoon, around 1992. It was one of those humid patches Ontario often gets in July, and I'd spent the entire day sweating on my hands and knees, weeding a strawberry patch.

"I was feeling grim, and even as a nineteen-year-old, my back ached. Whenever I closed my eyes, all I could see were strawberry plants. So, there I was, dragging myself in from the field to wash up, when my step-grandmother arrived back at the farm. My step-grandmother was a bit of an enigma. In some regards, she was typical of her generation — she was a school trustee, contributed to bake sales, that sort of thing — but she also ran an extremely successful investment firm and had a portfolio of wealthy clients who trusted her stock tips.

"Anyhow, on this day, as I was heading in from the field, her Lincoln Town Car, which was a deep aubergine color, sort of oozed down the driveway and she got out. I can feel — to this day — the puff of air-conditioned air escaping the car. She'd had a pretty good day, some big deal had come off, and as we chatted, I did some quick mental math. She had made more money that afternoon, just from commission on moving funds from one kind of investment to another, than I would make all summer, working full-time, producing food for people.

"I think that was the minute I decided not to take on the family farm. I loved that property and enjoyed working outside, but when decision-time came, and my grandparents offered me the farm, I talked it over with my wife and opted for grad school. Quite frankly, it is easier and better-paying to talk and write about food than it is producing fruits and vegetables on a family farm."

Lenore agreed. She has a similar story involving halibut and her family's fishing boats. Producing food, processing food, preparing food: these are hard jobs that generally don't pay very much.

Our stories are not unique. Over the past three generations, hundreds of millions of people have come to a similar conclusion. As the non-agricultural economy has expanded with manufacturing, service, and other jobs not connected with primary production (e.g., agriculture, forestry, and mining), folks have opted to spend their time doing things other than producing food, hewing wood, or hauling water. In fact, a major emphasis in international development right now is to help small-scale farmers develop labor-saving technologies, make them more efficient, free up their energies for more lucrative (or downright pleasant) activities, and reduce the drudgery of farm work.[1]

Mars is unlikely to be different. In fact, we expect labor on Mars to be pretty much like everything else on the Red Planet: extremely scarce. The price to put the first living, breathing human on Mars is estimated at five billion dollars, and even Elon Musk's optimistic plans for large-scale communities put the cost of a ticket to Mars at a half-million dollars. Hence, Martian labor will be at a premium, and early pioneers will be far too busy building biodomes, conducting science, and figuring out how to terraform Earth's second-closest neighbor to spend much time picking lettuce.

Agricultural labor shortages are nothing new. For centuries, agricultural technology has focused on allowing farmers to produce more food with less labor. Today is no different, and there are chronic farm labor shortages around the industrialized world as the domestic workforce in places like the U.S., the EU, Japan, and Canada can find better jobs than the long, hard, tedious work most farming demands. Lenore once explored the Japanese indoor vegetable production industry, and she was shocked to find most of the labor was being done by people in their eighties looking for a little part-time work. Without these grandparents, the industry would grind to a halt.

For farmers facing such a shortage, importing seasonal migrant workers is a tried-and-true strategy. In Canada, about sixty thousand

---

1   Evan's last book, *Uncertain Harvest*, has a chapter on this.

people arrive every summer from Latin America and the Caribbean to prune apple orchards, plant asparagus, and pick strawberries. But as we all discovered during the COVID-19 lockdowns of 2020/21, importing laborers isn't necessarily a resilient way to run a food system. And for the workers, it can be exploitative. These short-term immigrants too often find themselves in situations where their energies are abused while they have uncertain recourse. This international arbitrage, in which we bring labor to our fields or rely on crops grown in areas of the world with a lower cost of labor, is vulnerable to disruption. And even when things are functioning normally, moving labor and goods comes with a climate cost.

Another way of looking at this situation is in terms of what economists call opportunity costs. The human brain is the most powerful problem-solving device yet known, capable of creativity and ingenuity that eludes artificial intelligence. Having huge numbers of people do repetitive hand labor is a waste of this capacity. This is especially true for small-scale farmers — a huge number of whom are women — across the Global South. It's akin to using a massive supercomputer to do basic arithmetic and is a waste of human potential. On Mars, where each human will represent millions of dollars of investment, we will not squander human labor on mindless tasks. On Earth, we should think like a Martian and embrace alternatives to relying on human weeders, threshers, and winnowers. It seems that this future is already coming. One of the showstoppers at Canada's outdoor farm show in the fall of 2019 was billed as an introduction to a new era of agriculture, an era that will be autonomous, hands-free, laborless, environmentally friendly, and worked by robots. One company showcased at the event — called Dot, after the inventor's mother, Dorothy — is in the vanguard of taking the farmer out of the farm. Watching the Dot tractors move on their own through the fields was one part Terminator, one part Willy Wonka, with a sprinkling of the Little Red Hen. Looking at these machines move, the two of us realized that the future farmhand is likely to look more like the Mars rovers *Spirit* and *Opportunity* rather than any version of Old MacDonald.

The Dot tractor demonstrated the promise of self-driving cars, and watching their ballet gave the audience a glimpse of what it must have been like a little over a hundred years ago as pioneering farmers unhitched their wagons to take in a demonstration of the early steam-powered farm equipment. Did the blacksmiths, farriers, and horse breeders of that generation realize that the winds of technology had shifted and theirs were industries destined to decline? Is the same true today?

The technology on display that afternoon in Ontario seemed both futuristic and retro. There was grace and agility as the machines moved up to each other, and with some hissing hydraulics and clicking joints, coupled and deployed into the field where they followed a path predetermined by an impressive array of soil analysis, remote sensing, and other forms of data. And if the machines ever became stuck, or when they met unexpected obstacles, these massive robots simply stopped moving and sent a message via mobile device so the operator could hop in the pickup, head to the field, and check out what was wrong.

One of the promotional videos used to advertise the Dot tractors suggests that this new generation of farm-bot will free farmers from the tedium of sitting all day in the cab while another suggests that one of the benefits of the technology is to allow robots to do anything that is boring, repetitive, or even dangerous. With this freedom, the farmers will probably do what most of us would do when freed up from a job — they will simply engage in whatever other pursuits a person might like to do if they didn't have to drive repetitively around a massive field for hours at a time. Both Martian communities and terrestrial farms will undoubtedly boast more robots than people. Farmers will tend to the human side of the business, and the machines will handle everything else.

MORE FOOD; LESS POLLUTION

It's not just the farmer who will reap the rewards of this upcoming digital agricultural revolution; the environment should benefit as well. This is because it's not just the tractors that are changing, but the tractors represent a complex ecosystem of technological innovations that include artificial intelligence, robotics, the Internet, satellites, and drones. With this technology, farmers can be more precise with how they plant and apply fertilizer and pesticides, and more precision means less waste, higher profits, and less pollution.

In the 1980s and 90s, when Evan was working on his grandfather's farm,[2] he spent a lot of time standing on the back of the family's old red Massey Ferguson tractor while his grandfather drove slowly through the fields of sweetcorn, melon, or strawberries. One of Evan's jobs was to throw handfuls of fertilizer pellets off the back of the tractor and onto the crop. With the benefit of hindsight, and a Ph.D. in environmental studies, Evan is quite certain now that at least half of this fertilizer did not land at the right place and time for the plants to use. And so, these valuable nutrients ended up as pollution — either in the atmosphere as a greenhouse gas or washed downstream into Lake Erie where the fertilizer would have fed the periodic blooms of "algae" (actually cyanobacteria) that are destroying the ecology of the lake.

And while to this day Evan feels guilty about how profligate he was when fertilizing his grandfather's fields, this story isn't uncommon. Across the world, fertilizer is overapplied and becomes pollution. This is one reason that agriculture today is the world's largest source of water pollution and responsible for about one-third

---

2   And apologies to any reader who has already heard this – Evan really does overuse this story.

of all our greenhouse gas emissions.[3] We need to do better, and one approach is through sustainable intensification; in other words, we must devise ways to ensure that we maximize food production while minimizing the harm. Lenore has watched the same dynamics unfold on the small farms of the Fraser Valley on the Pacific coast, where accumulated runoff threatens the beloved salmon runs. This kind of pollution is found nearly everywhere we farm.

The logic of sustainable intensification on Mars is irrefutable. Since there is no topsoil or water to exploit, Martian growers will have to make every possible resource count. But we struggle to get this same efficiency on Earth because there is no simple formula for sustainable intensification. The ways a farmer would sustainably intensify a U.S. Midwest corn and soy farm will be different from how a farmer would sustainably intensify a small-scale rice farm in rural Asia. And it's on those massive North American grain farms that the iFarm and the Dot tractors are most critical. Unlike teenage grandsons chucking handfuls of nitrogen pellets randomly off the back of a tractor, Dot's smart tractors "know" where they are in the field and use this knowledge to plant the right seed in the right place and give it the right amount of fertilizer. The smart tractor, therefore, will be able to maximize food production while reducing energy use, costs, and pollution.

Lenore remembers a different set of problems but a similar challenge; each time a Newman boat left port on Canada's West Coast, there were hours of grinding tasks to perform. Miles of line had to be inspected and damaged hooks replaced. This was slow work that caused repetitive strain injuries in her family. Bait had to be cut into cubes, and hundreds of handmade wood and metal parts had to be serviced. Over the years, some of these tasks

3   Gilbert, N. (2012). One-third of our greenhouse gas emissions come from agriculture. *Nature*. https://doi.org/10.1038/nature.2012.11708

were automated. In some cases, frustrated fishermen devised their own technology and commercialized it. Every step forward in the fishing industry has resulted in savings through the automation of repetitive tasks and the standardizing of equipment.[4]

Evan called Lenore to bring her up to speed on this research.

"To really drive a full wave of automation, as well as to allow that first Martian community to thrive, I think that the agricultural robot of the near future will use a suite of novel technologies. First are the robotics, artificial intelligence, and sensors that are similar to self-driving cars. In addition, these tractors have what farmers call variable application technologies that can do different things in different spaces on the farm."

Lenore cut in, "So instead of a farmer carting around a single tank of spray or a single bin of seeds, an iFarmer's equipment will sport a range of different spray tanks (or seed hoppers) and so be able to plant different seeds in different locations? Is that what you mean?"

"Exactly," replied Evan, "And all of this will be made possible because the smart tractor 'sees' the farmers' fields differently from the farmer."

When the typical industrial farmer sees a field, she or he mostly thinks of it as a large space to be planted, tilled, and fertilized in a uniform fashion. Cornfields, for example, are planted and harvested as units, and only rarely do you see commercial farmers do different things *within* the same field. To the farmer of the late

---

4   Sadly no one ever figured out an easy way to paint the fifty-two-foot-long boat, *Jaana*, a yearly task that took multiple family members days of backbreaking work. Lenore loves the ocean but is happy her current job takes place away from the corrosive effect of salt water.

twentieth century, large-scale uniformity was a friend. But these iFarm robots, epitomized by Dot's smart tractors, perceive fields as being composed of different management units. For instance, highly productive management units would be areas within a field that contain rich soil and predictably produce a bumper crop. Other areas would be identified as having lower potential and, therefore, needing a different variety of seed or quantity of fertilizer to optimize productivity. Taken together, the robotics, the sensors, and the variable application technologies allow artificial intelligence algorithms to decide what to do on a meter-by-meter basis. In short, the robots can do something humans can't do: provide tailored attention to every part of a field and in doing so plant the right seed in the right place and give it the right amount of fertilizer. [5]

To be able to do this, the iFarm's smart tractors need data. Lots of data. That data will need to be collected by smart harvesters over time to build up a detailed field map showing where fields are more or less productive. As well, farmers might collect data by flying drones over their fields to provide insights into things like pest outbreaks or signs of drought. Remote sensing and satellite data provide detailed maps on the size and shape of fields while more detailed ground-level scanning reveals higher or lower spots, poorly drained areas, gullies, trees, or other obstacles. All these data could be linked through to

---

5   Anyone who spent part of 2020's lockdown in the backyard garden has experienced this phenomenon. Lenore and her father used some of the time during the lockdown to plant a big kitchen garden and produced a tasty calendar of local produce from rhubarb to winter chard. But in doing so they noticed a patch of soil of about a meter square that just wouldn't produce. Seeds planted in this area didn't sprout, and transplants died. All fields have spots like this, and robots might be the key to unlocking all of a farm's potential. Lenore's father solved the problem of the garden's "Bermuda rectangle" by parking a lawn chair on it, not an approach that works on the scale of a commercial farm.

market information on the price of the hoped-for commodities as well as their attendant costs such as the price of fuel and fertilizer.

iFarm's onboard analytics, backed up with massive amounts of cloud computing power, will help smart tractors decide what to plant, how much fertilizer to apply, and when to harvest, maximizing productivity and minimizing waste.

On Mars, gathering all this data will be easy because all the farming will be conducted inside. Drones are getting smaller and more agile, necessary when navigating inside Martian domes. On Earth, connecting this technology to the farmer is more difficult, but the effort to colonize Mars is coming into play on this front as well. Elon Musk's company SpaceX is, right now, weaving a net of four thousand low orbit satellites around Earth to create a network called Starlink that will provide Internet to all parts of the planet, including to the many rural farmers not yet served by reliable access. Nearly one thousand of these satellites are already in orbit, visible as they cross the night sky above Lenore's garden. In fact, SpaceX plans to put a similar network around the Red Planet, meaning that the first generation of Martian farmers won't have to struggle with connectivity. All of which brings us back to the reasons for this technology in the first place — if properly deployed, the tools of the Digital Agricultural Revolution should allow us to be more efficient with the resources that go into producing, processing, and distributing our daily bread. These farm robots give farmers — and especially large farmers in places with lots of land — new tools to be more precise. The data-powered iFarm is ringing in a new age of farming that helps both the environment and the producer.

Or at least we hope it works this way; but before we get there, several kinks still need to be sorted out.

IT'S THE SYSTEM, STUPID

A few years ago, Evan had an opportunity to hear one of the greats of agricultural economics, David Pannell, a soft-spoken academic from the University of Western Australia who has spent a lifetime studying farmer behavior. As the director for a research center devoted to understanding environmental economics and policy, Pannell is the author of hundreds of journal articles and books. His job is difficult; the ancient red soils of Australia are notoriously fragile.

In a dimly lit conference facility, Pannell stepped the audience through the reality of the digital agricultural revolution, helping to cut through some of the marketing hype. He did this in a way that made esoteric economic concepts exciting, explaining what economists call flat payoff functions. For example, when a farmer applies fertilizer or irrigation, they expect to see their profit rise, if they keep applying them, then profits will plateau before ultimately declining.

Pannell's evidence shows that across a wide range of farms, and a wide range of inputs, there are some common features to the way crops behave when farmers apply things like fertilizer. In particular, he showed that while there is an optimal level of fertilizer on every given field, the amount of money a farmer earns from a crop doesn't vary that much whether they over- or under-apply it. Hence the idea of the flat payoff function. In other words, the payoff for adding fertilizers or other inputs lies within a narrow range and is not very sensitive to the amount farmers apply. In fact, Dr. Pannell showed that a farmer is more likely to lose money by *under*-applying fertilizer than they are if they *over*-apply it.[6]

---

6   https://www.coursera.org/lecture/agriculture-economics-nature/video-flat-payoff-functions-Nw8uu

The good professor concluded that not only are these flat payoff functions very common, but they demonstrate that a farmer doesn't need to be particularly precise with how much fertilizer they apply to receive the same level of profit. All of which leads to a question: even if the iFarm is good for the environment, is it worth it to farmers to invest?

On Mars, the economics of agriculture will be even more cutthroat. Any input a Martian uses must be manufactured at the expense of something else in the community's very tight industrial ecosystem. Will some crops simply be too resource-intensive? Maybe, but because getting these balances right means survival of the community, Martians will pore over any available data and might decide to focus on fewer, less resource-intensive crops. But bringing such decisions to Earth might require more subtle instruments.

What we learn about the application of digital agricultural technologies provides an important lesson about how we apply technology in general. Namely, while technologies may enable solutions to things like environmental problems, they rarely are the solution in and of themselves. In other words, no matter what the theoretical benefits or potential of a technology, technology needs to succeed in the real world and solve real-world problems. This is particularly important when it comes to the environment. If one of the hoped-for benefits of the iFarm on Earth is to reduce agriculture's environmental footprint, and if we want farmers to use the technology for this purpose, then we had better work with government to devise an incentive program that provides a financial reason for farmers to do it. In this, farming is no different from any other sector of the economy, and unless the policies, financial incentives, and regulatory frameworks are there, it is naïve to assume that farmers will shoulder this expense voluntarily. The technologies are not a panacea in themselves.

A MURDER (?), HERD (?), FLOCK (?), SCHOOL (?), OR SWARM (?) OF TRACTORS

"Do you think that Martian tractors will look like the Dot machines?" asked Lenore one day after Evan shared the YouTube videos of the machines waltzing autonomously through the fields.

"To be honest, I don't think so," Evan replied. "Despite the marketing material, the kind of smart tractors epitomized by Dot's technology may not be a good fit for Mars. First, they are very large pieces of equipment and only suited to farming big areas that have uniform fields. That's not Mars and that's not an accurate description of farming on many parts of Earth either. Second, they are mostly used for crops such as grains and oilseeds. And as we've already discussed, these aren't the crops that are likely to be all that common on Mars. But even though the Dot example might not be a good fit, we could look to parts of the world where small-scale intensive farming still exists and see how farmers in those regions are using the same underlying technologies."

When it comes to exploring disruptive technologies, it seems fitting to turn our attention to Google. The company that once disrupted the Internet is now throwing its almost infinite weight behind their X company, which is devoted to finding moonshot solutions to pressing global problems. One of X's projects is tantalizingly titled Mineral and posits that to meet the needs of the growing human population, we need to change our food systems. In their promotional material, the minds behind Mineral make an interesting point: the basic agricultural approach, common for about ten thousand years, has been to standardize what we grow and how we grow it. This means that today in North America a tiny handful of plant and animal species dominate the entire continent. Chickens, cows, and pigs, along with wheat, soybeans, and corn, are all produced using near-identical management practices and stretch more or less without interruption from the Rockies to the Atlantic and from the boreal forest down to the Mississippi Delta.

Those Green Revolution technologies we described earlier, have reinforced this trend to bigger farms, fewer crop varieties, and standardized farming practices. The hybridized seeds, irrigation, pesticides, and synthetic fertilizers that were invented almost a hundred years ago have been so successful in boosting harvests that they dominate farming systems to a huge degree. This trend might feel unstoppable, but it is not. The next revolution is on its way and is headed in a radically different direction.

It seems likely that the new technologies of AI and robotics, rather than forcing us to ever-larger economies of scale, give us the potential to farm at small scales. The webpage for Google's Mineral puts it like this: "What if new technologies could help us embrace nature's diversity and complexity, instead of simplifying it? If growers could understand how each plant on their farm is growing and interacting with its environment, they could reduce the use of fertilizers . . ."[7] And so the smart women and men at Google are doing things like putting sensors on bicycles that are moved slowly through small fields and dreaming of ways to monitor every single plant in a field.

"What I've learned," said Evan, "is that the real paradigm shift driven by digital technologies is to allow us to do things at the small-scale — not nano-or-micro small (as discussed earlier), but at the scale of the individual plant or animal. The roots of this change are less likely to come out of North America's grain fields where everything is about bigger and bigger but in places where

---

7   https://x.company/projects/mineral/

small-scale still matters. And thinking at this scale takes us to other places in the world where the same suite of technologies is breeding an entirely new generation of farm equipment.

"I think that the place on Earth most analogous to Martian food systems may be in Africa or Asia, where things are still done on small pieces of land. It's even possible that the kinds of digital tools I've been reading about may emancipate women, reduce drudgery, and drive down poverty as well as improve the sustainability of farming."

Lenore thinks Evan may be onto something here. In one of our conversations, she noted that one reason we depend on so much hand labor in farming — and exploit women and female-headed households to do the worst of it — is partly because our capitalist economic system does a bad job of valuing their contributions. But on Mars, labor will be scarce, and every worker will represent a huge investment, so we won't dump all the crappy jobs on systemically mistreated groups — we'll get machines to do it instead.

And using technologies like AI and big data to improve small-scale farming will be an ally. Some companies are investing in what are affectionately known as *tractor swarms*. A tractor swarm is a flock of small autonomous vehicles all networked together but able to act autonomously. Imagine a group of wastebasket-sized robots on wheels, each equipped with a single cultivator, seed drill, or sprayer. These all spread out individually into a field and proceed to scurry around, tending plants, pulling weeds, seeding, or spraying as needed. This is perfect for an acre of farmland on the edge of a major city or for a compact greenhouse drilled into the surface of another planet.

We are seeing evidence of companies with tractor swarms and drone flocks springing up in places such as Cambodia where farmers can book drones to help shift small loads in or out of a field. One of our favorite examples is a research project called MARS, which is short for Mobile Agricultural Robot Swarms. MARS uses drones to scan fields and create detailed digital maps, and these data are used to coordinate the autonomous mobile robots when they are in the field. Between drones and the on-the-ground robots, the farmer has access to a wide range of flexible tools that can either grow or shrink in scale depending on the size of the farm or the task at hand. MARS bills itself as a boon for small-scale farmers who want to reduce pesticides, and it presents itself as a tool for organic farmers and those operating on relatively small plots of land.

The recent invention of the tractor swarm redefines the very notion of what a tractor is. No longer must a tractor be a massive capital investment available only to the largest of farmers. No longer is buying a tractor part of an economic package that leads inexorably to greater farm consolidation and eye-watering economies of scale. Instead, in the very near future, a tractor may be a fully scalable network of small vehicles and drones driven from a mobile device and accessible to a diverse range of farmers.

And just as the very notion of the tractor is being redefined by the technology, so, too, is the notion of owning a tractor. Uber-like companies are also starting up in places such as Africa and South Asia, offering farmers the ability to rent the services of small-scale equipment on an as-needed basis.

This means that farmers who struggle to put capital together to make a major purchase could join up with their neighbors and get some much-needed mechanical help for a period, thus easing the drudgery of their work and improving their productivity. Imagine a group of rice farmers in India who collectively order the services of a delivery drone to bring their harvest to a threshing facility, or a collective of female farmers in Ghana who have their fields cultivated at the same time using the services of an entrepreneurial start-up that rents tractor swarms. When considered in this light, the applications unlocked by artificial intelligence, autonomous devices, and data analytics allow us to radically redefine the way that farming is done, and this represents a truly disruptive approach to technology.

The real beauty of the iFarm, therefore, isn't the showstoppers at the big agriculture fairs in North America. These autonomous behemoths will likely play a role on farms already dominated by massive economies of scale. Rather, the disruptive innovation will occur when the technologies are packaged to work at a much smaller scale. By allowing farmers to manage crops or livestock at a small scale, and in real time, we'll see the technologies emerge that are suited to supporting the food system on Mars.

THE RIGHT STUFF

On Mars, every plant will matter and will require detailed monitoring and care. The systems providing that care will need to be as skilled as the human Martians who guide them.

According to NASA, to be an astronaut, you need to have training in the so-called STEM disciplines of science, technology, engineering, or math. Or you can be a doctor or test pilot. Then you need a fair bit of professional experience where you have demonstrated "skills in leadership, teamwork, and communications," and you need to pass a physical endurance test (oh, and be a U.S. citizen, but that bit is less relevant for the purposes of this book).[8]

The skills website careergirls.org breaks it down even further. They suggest a would-be astronaut needs analytic, communication, listening, decision-making, mechanical, math, and teamwork skills. Would-be Martian community scouting agents would want similar

---

8   https://www.nasa.gov/audience/forstudents/postsecondary/features/F_
     Astronaut_Requirements.html

skills — to survive in space you need unbelievable grit and determination along with superb interpersonal skills and high-level training in some technical discipline. As crazy as it sounds, both Evan and Lenore know that this is also pretty much what the ag-tech sector says we need in our future workforce if we want to ignite the future of food. Getting future farmers Mars-ready involves a lot of the same technical skills that agricultural workers will need on Earth.

Today, the traditional agricultural science disciplines such as crop or soil science are as important as they ever were. And this means that the would-be agri-preneur of tomorrow needs to understand how plants grow, how to keep food safe, and how to maintain soil fertility. As well, the arriving wave of agri-tech innovation will also require a keen awareness of the STEM subjects (science, technology, engineering, and math). The problem is we don't really do a very good job teaching data science or coding in our agricultural colleges. To address this gap, agricultural faculties and colleges need to do a better job of exposing their graduates to these things. Equally, our engineering and computer science schools need to embrace agri-food as a source of case studies, job placements, and co-op work experiences for their students. No longer can agriculture be seen as a quaint rural activity framed by a straw hat and a red barn. Producing enough nutritious food for the world's billions while protecting planet Earth deserves to be seen in the same light as aerospace, automotive, medicine, or IT.

But even merging STEM courses with traditional agricultural disciplines won't be enough. For instance, our agricultural schools typically focus most of their energy on conventional agriculture, meaning our students mostly learn how to produce mainstream commodities and livestock at large scales using Green Revolution technologies. Agricultural educators need to broaden this and explicitly include alternative production systems including regenerative agriculture, biodynamic agriculture, organic agriculture, and the gamut of high-tech innovation described throughout this book.

Equally, workplace studies reveal that to thrive in the new world of work, young people need to be excellent technicians, strong in the scientific disciplines, and grounded in the so-called soft or foundational skills of active listening, critical thinking, oral and written communications, and project management. In this regard, our current ways of doing things in the typical university setting, where lectures, exams, and essays still dominate, lets students down. Most of the training most students receive in most university programs is still based on a pedagogical model where students acquire information from a professor and then display that they have mastery over that subject by, on their own, writing an exam or paper. But in the real world, teamwork, adaptability, and project management are just as important as technical mastery. On Mars, and on Earth, farms will be connected through an industrial metabolism with wastes and inputs circling through the whole community. Farmers will need to be team players.

A few years ago, Evan was asked to participate in a launch event for a report on the future of work in Canada. The report was published by Canada's largest bank, the Royal Bank of Canada, and makes many of the points we've articulated in this chapter. Namely, to thrive, students need to be great at technical disciplines but also adept at the foundational skills and our current university and college system needs to change quickly.

At the event, Evan found himself on stage with the head vintner for a large Canadian winery[9] who demonstrated how she could use her smartphone to control the irrigation and wind machines in the winery a hundred miles away. She then admitted that she didn't

---

9   That the vintner was a woman in her thirties already shows how the sector is changing.

need to do even that much because sensors, located in the soil and on the plants, measured everything from moisture to temperature and controlled the irrigation and the wind machines autonomously.

The real kicker, however, was when she said that although the technology has allowed her organization to reduce the number of people working in the field, the winery had increased its payroll by hiring more data technicians and software engineers, sommeliers, marketing gurus, and a whole host of people with other skill sets. For Evan, this moment neatly illustrated both the challenge and excitement presented by the digital agriculture revolution.

Here is what Lenore and Evan think is likely on Mars. Everyone will be a scientist at heart. Sensors that monitor all aspects of crop life will generate huge amounts of data that will drive small fleets of autonomous vehicles to do most of the habitats' planting, tending, and harvesting of food. If the systems break down, human technicians troubleshoot, but these humans will work in teams and the management of the growing systems will only be part of their portfolio of responsibilities. The human technicians, who will spend most of their time working on non-food-related activities, will need highly developed technical skills plus the interpersonal skills that make for great team players able to get along in a tough setting.

Pretty much the same can be said for the future of food systems on Earth. To deploy new technologies, regardless of the planet, we need a new generation of farmers to be trained in a way that is radically different from the past. To feed the future, the farmers of tomorrow need to be trained as if they are going to become astronauts. They will truly need "the right stuff."[10]

---

10 Ok, we are probably dating ourselves with this reference, but for readers not familiar with obscure movies from the 1980s, *The Right Stuff* was a Hollywood flick about astronauts.

CHAPTER 9:

Closed Loops

NATURE'S ELEGANT SOLUTION

There is a game that Evan, ever the sci-fi fan, plays most years with his graduate student classes. It's called the *Spaceship Exercise*, and in it, Evan asks students to imagine a generation ship, a spaceship designed to keep a fragile human cargo alive for many decades on a hypothetical voyage between the stars. The students are tasked with establishing the basic principles by which this generation ship needs to operate to ensure that the descendants of the original crew arrive at their destination in good health.[1]

After overcoming the initial shyness at the absurdity of this thought experiment, Evan has observed many cohorts of grad students get into the exercise with enthusiasm.

---

[1] By contrast, getting to Mars should only take between six and eight months. So, considering what's needed for a ship that would have to keep folks alive on an interstellar voyage truly pushes the envelope.

"A steady supply of food," someone always shouts early in the process.

"Water, and air to breathe."

[. . . such basics are always the first ideas put on the table. Then, things usually shift to become a little more technical . . .]

"Protection from harmful radiation."

"A constant temperature."

[. . . before veering off into more social realms . . .]

"Good quality birth control."

"Social activities and goals to keep away the tedium."

"A way of expressing culture and art."

"A research facility to foster ongoing innovation."

It's never long into the exercise before someone makes the point that to achieve these laudable aims, the space voyagers need a ship with extremely sophisticated systems to recycle pretty much everything. When you're in space, you don't have the luxury of digging up new resources when you run low on something, and you certainly can't let stuff lie around and accumulate until it becomes pollution. At this stage in the exercise, the idea that we are talking about a spaceship has vanished and everybody is openly discussing what we need to do, here on Earth, to ensure the long-term sustainability of our planet.

When Evan asked Lenore the same question, she replied immediately.

"Energy sources and entropy sinks. Lots of entropy sinks . . ."

Lenore knows that such answers are one of the reasons why people don't ask physicists questions very often. But she also knows that perfect recycling is only possible with a lot of energy and a place to dump the heat and waste products. The concept of entropy, though, isn't needed to understand the general idea. It is much easier to maintain a functioning life-support system on a large planet with a biosphere than it is to maintain such a system on a Martian base, space station, or spaceship. She also knows that when her father took the family vessel, the *Jaana*, out into the middle of the Pacific to fish tuna, he took multiple copies of critical parts and the tools needed to make more parts. The principle is roughly the same.

Evan has been thinking about the importance of recycling in food systems for a long time. When he was a Ph.D. student in the 1990s, he remembers going to a lecture by the renowned novelist, essayist, poet, and philosopher Wendell Berry. Berry used the lecture to reflect on civilization and the world by thinking about what he called "nature's elegant solution," by which he meant the cycles of nature where there is no such thing as waste.

Berry reminded Evan of what most schoolchildren know, and what the students doing Evan's spaceship exercise always deduce: nature works only when it works in loops. The tree drops its leaves, which decompose, enriching the soil, before they become part of the tree again in subsequent years. The problem is that long ago, humans began breaking the tidy cycles that contribute to the elegant solution, thereby creating what Berry called two wicked problems. At one end of the economy, we destructively extract resources from the environment, while at the other, we struggle to dispose of waste. The deforestation and soil degradation driven by our farms, forestry, and mining operations are matched by the food waste and plastic pollution spilling out of our cities, choking oceans and waterways. Unfettered, the Industrial Revolution has exacerbated our impact with more and more of us living in a take-make-waste economy. But if we want to colonize Mars, or simply live more sustainably here on Earth, we need to do better. We are going to have to close the loop on our food systems and dust off Berry's elegant solution. If we don't, it's unlikely that *Spaceship Planet Earth* will survive long enough to give us a shot at the stars.

PROBLEMS WITH THE INDUSTRIAL REVOLUTION

To explore how to create a circular economy of food, we need to step back and unpack some of the economic foundations of the modern world. The notion that we all live in an industrialized economy owes its roots to ideas that arose more than two hundred years ago when folks such as Adam Smith did some of the pioneering work in the emerging discipline of economics. As anyone with a high school understanding of economics knows, the prices consumers pay for a good or a service are supposed to be determined by the supply and the demand of that good or service, the costs of producing those goods and services, plus a bit of profit to sustain those who did the hard work. Done right, markets result in an efficient allocation of resources and create rewards for innovative people who bring new ideas to consumers.

But things don't always work quite so neatly.

We have problems such as pollution, that represent very real costs to society but are not always captured by the market. For instance, no one is really paying for the long-term costs of greenhouse gases, yet we know that climate change will make life much more challenging in the near future. By not assigning a meaningful cost to greenhouse gas emissions, it's easy to ignore them in the short-term. But there is a newer generation of economists who argue that governments have a role to play in correcting for these market failures and should put in place policies such as carbon taxes. Digging deeper, however, there are other problems with the way that markets work. There are two that are particularly concerning.

The first is called the theory of comparative advantage, basically an academic way of saying that costs go down if firms specialize in producing only those things that they are good at. When it comes to food, the theory of comparative advantage generally translates into something like the following. Let's imagine two countries — Camelot, which produces excellent tomatoes because it enjoys hot summers and a landscape that lends itself to smaller fields; and Xanadu, which has bigger fields and is naturally suited to wheat. If Camelot focuses on growing tomatoes and trades with Xanadu for their wheat, then consumers in both will be able to enjoy cheaper pizza than if farmers in both Xanadu and Camelot had tried to produce both products themselves.[2] The second theory from economics has a similar flavor but applies to labor. This is called the division of labor and suggests firms can reduce costs and become more efficient if labor is allowed to specialize. As Henry Ford's assembly line shows, it is efficient to have workers specialize in a narrow range of tasks rather than asking everyone to do everything.

When applied in the real world, the logic of both comparative advantage and the specialization of labor really do make things pretty efficient. And these theories mean that modern food systems have become incredibly specialized with huge regions devoted to only producing a small number of commodities such as wheat, canola, and lentils in the Canadian prairies or corn and soy in the U.S. Midwest. As such, the application of these two theories has pushed the modern world straight into the arms of Wendell Berry's "two wicked problems."

On Earth, as well as for our hypothetical Martian community, we cannot afford this system any longer and we must create closed-loop or circular food systems where every by-product finds a useful

---

2   Technically, what we are describing here is the theory of absolute advantage, where one country has an absolute advantage in producing something over another country. Comparative advantage is a variation on the theme and means that even if one region can do better at producing both products, everyone is still better off if they specialize on producing (and then trading) what they can do best comparatively speaking.

home somewhere else in the system. At the same time, we must work to achieve the kind of efficiencies that can only be gained through divisions of labor and comparative advantages. Given that one of the negative consequences of the Industrial Revolution was — in the name of efficiency — to cleave nature's elegant solution into the twin problems of resource depletion and nutrient surplus, is it possible to design a system that is both economically efficient and closed-loop here on Earth as well as in space?

FLYING IN EVEN TIGHTER CIRCLES

Jurek Kolasa belongs to a breed of scientist, engineer, and entrepreneur committed to finding a closed-loop solution to the industrial food system's linear problem. From his base at McMaster University, in Ontario, Canada, he creates modular designer ecosystems in large tanks. In doing this, Jurek hopes to better understand the Earth's ecosystems, improve food security, generate jobs, and save the environment. Jurek was speaking to Evan from his home in Hamilton, near the shores of Lake Ontario.

Even across the Zoom link, Jurek's enthusiasm for his vision was evident as he explained to Evan how the secret to developing closed-loop food systems is to think of the ways natural ecosystems work and where each organism plays a part in an integrated whole. Some creatures, such as the cyanobacteria we met earlier and their cousins the algae, are autotrophs, or primary producers, that harness sunlight and pull nutrients from the water. Other organisms exist by eating nutrients, thus preventing pollution. Still others (grazers) keep the populations of the primary producers in check and together the community of organisms make up a tidy whole. It's the basic web of life we all learned about in grade school, but Jurek's vision is that each of these different functions can be isolated and engineered into closed-loop systems where the product of each stage becomes an input into the next step. This is how nature works in the wild, and Jurek's vision is that this is how our food systems should be designed on Earth and on Mars.

When pressed to explain his idea, he describes a series of tanks or containers, each with temperature and pressure controls. In the first tank, algae or other microorganisms grow and provide food for tiny critters in tank two, the tiny critters feed bigger critters in tank three, and so on until the final tank produces large (and tasty) fish suitable for human consumption, while the waste products from the fish are recycled back into the first tank along with the water. The humans managing the system harvest a surplus from each tank, recycle everything else, and keep the whole system going by adding a few nutrients, sunlight, and a small amount of water as needed. It's not completely self-sufficient, but once running, such systems require an incredibly small number of inputs compared to the industrial food system we currently depend on.

"The key," Jurek said to Evan, "is to have a diversity of organisms in each stage, and every one of these organisms must fulfill different functions. So, yes, you'll want your cyanobacteria in some of the early tanks, but if you depend too much on a single species to do key functions, and something goes wrong with that species, then the whole system may be affected. On Mars, that could be catastrophic."

Jurek suggested, therefore, that rather than thinking about single species, communities of species should be harnessed to create a life-support system. He then pointed out that along with cyanobacteria, spirulina and other forms of algae would be helpful.

He laughed and said, "If you overspecialize, things are a bit risky as you have to make sure everything is just right for a very small number of organisms. However, if you have a community made up of different kinds of algae, then it's likely the overall system can tolerate a wider range of conditions, and this makes things more resilient."

"And on Mars?" Asked Evan. "This will be necessary on Mars?"

"Absolutely," came Jurek's reply. Then he looked directly at Evan and said, "But you do realize that it will be much easier and cheaper just to save Earth than go to Mars to figure these things out only to return to Earth?"

Evan nodded and Jurek continued, "A Martian community is a great thought experiment, but I think we should focus on Earth first, no?"

Despite Jurek's enthusiasm, and the theoretical elegance of closed-loop systems, the reality of implementing a closed-loop paradigm in a commercially viable way seems daunting and brings us back to the power of Adam Smith's theories around comparative advantage and division of labor. To explore this in a little more detail, it's worth digging into the economics of a special kind of greenhouse that really tries to embody much of what Jurek was talking about. Called aquaponics, these operations attempt to mimic closed-loop systems by bringing fish and vegetable production together under one roof.

A high-level aquaponics operation involves keeping fish in pens or tanks next to greenhouses and using the waste from the fish as the fertilizer for the vegetable crop in the greenhouse. This is basically a simplified version of Jurek's ambitious vision, and the greenhouse trade magazine *Ceres* suggests that the modern aquaponics operation is a window into a future that uses "sustainable growing methods, creates zero waste, and gives back to the community by supplying fresh, local food."[3]

But while this sounds good in theory, there are big challenges to implementing these systems. First, anything smaller than a ten

---

3   https://ceresgs.com/guide-planning-commercial-aquaponics-greenhouse/

thousand square foot operation wouldn't really be commercially viable given the very high upfront costs. Such a huge operation would likely be owned by investors, heavily dependent on expensive automation, and run by hired managers and workers. The problem is that such an outfit would have to compete with both the traditional greenhouse sector and specialized fish farms. But because the aquaponics operator is trying to do two things (produce both fish and veggies) they typically are not as efficient as the businesses that specialize in doing just one thing well. So, these sort of big aquaponics operations are quite rare. As a result, most of the entrepreneurial aquaponics start-ups are small (i.e., less than ten thousand square feet) and despite great intentions and a compelling business case, these small-scale operations don't seem to last long.

For instance, a few years ago Evan did some work with two different aquaponics companies based in or around Toronto, Canada. One of these had even been named Ontario's "social enterprise of the year" in 2018 for being devoted to principles of sustainability. But when Evan reached out to his connections, he discovered that they, like many others, had simply vanished and gone out of business.

Frustrated with the lack of progress, we regrouped to debate aquaponics' future. The tone of our chat was more muted than usual, but Lenore immediately identified a couple of possible problems.

"First," she said, "it's that question of scale. A lot of these aquaponics companies have never really been able to compete."

Producing food is a low-margin business, and if you're going to compete in an already crowded marketplace, you either need to convince the consumer to pay a premium or you need to have sufficient volume to go head-to-head with conventional agriculture.

"Next," Lenore said, "it's a question of capital."

Some of the aquaponics start-ups Evan knew a few years ago did quite well by Canadian standards and managed to raise a few million dollars. But a few million is just a drop in the bucket for a food-tech start-up. In fact, they need hundreds of millions of dollars in venture capital to get up and running. Think of Uber or Twitter, both of which spent years hemorrhaging money before ever showing a profit. These tech giants were propped up by wealthy investors who bought the logic that eventually these companies would make stupendous amounts of money and completely disrupt entire sectors of the economy. While the venture capital world understands that this is the case for transportation and IT, the same funds have not yet flown into the food sector. When it comes to investing, aquaponics players are small fish in big ponds.

Finally, aquaponics operations — and food production in general — are a risky proposition in part because they are working with living systems that are inherently harder to control than (say) a software environment. Also, the aquaponics sector is especially tricky since aquaponics companies are fundamentally trying to produce two things: fish and vegetables. Because of this, these systems are much more complicated to run and have a lot more things that could go wrong. Consequently, they end up spreading the resources and expertise of their teams a bit thin. Harkening back to Adam Smith's logic about comparative advantages and divisions of labor, and despite the compelling environmental logic of closed-loop systems, it is possible that the aquaponics business model is flawed in that it doesn't allow entrepreneurs to become sufficiently specialized that they can be efficient.

As the two of us debated these points, it seemed we had reached a point where we were beginning to disagree. For pretty much all the issues covered in this book, the two of us found ourselves in almost perfect alignment. But on the issue of aquaponics, it seemed that our consensus was starting to fray. Lenore held out hope that there was a solution to this problem while Evan, quite frankly, was ready to toss the entire idea of aquaponics into the trashcan of good ideas on paper that don't work in the real world.

Or, at least, he was until he and Lenore had one further interview that got both of us thinking about this whole topic in a new way.

OUR FOOD FUTURE

Cher Mereweather has an infectious energy for all things food-related and spoke to us from her home in the Wellington region, on the outskirts of Toronto. With a career in both civil society and the food industry, Cher is that rare person who is known and liked by everyone she meets. She is a consummate connector and dealmaker, a broker of relationships, and a force of good in her home community.

Evan, Lenore, and Cher met over Zoom around the winter solstice of 2020, as the shadows were gathering outside and everyone was hunkering down for a second bout of pandemic lockdown. But despite the heaviness of the year, Cher was pure energy.

"Everyone says what a drag 2020 was, but quite frankly, 2019 was worse." She said this earnestly, and while this initially seemed crazy to Lenore and Evan, as Cher explained, her comments started making sense.

Cher is the CEO of an organization called Provision Coalition whose purpose is to help companies make food sustainably. However, in 2019, a major reorganization of governmental budgets had cut Provision Coalition adrift from the public purse and forced it to fend for itself. But at exactly this time, a movement was brewing across the region to do something about food waste. Guelph city hall had teamed up with Wellington regional government to raise ten million dollars from the federal government's Smart City Challenge and launch a project called Our Food Future that is designed to create Canada's first circular economy of food. The overarching goal of Our Food Future is to try to rebuild that elegant solution Wendell Berry introduced to Evan decades ago.

The project Cher told Evan and Lenore about — called RePURPOSE — flowed out of that work and is designed to show that food systems can be economically viable and nutritious, yet not result in excessive resource extraction and waste. As Cher told us about this, she laughed, because the story of RePURPOSE begins, as many good stories do, with beer.

On the edge of the entirely nondescript midsize Canadian city of Guelph, adjacent to an entirely nondescript industrial park, off an entirely nondescript stretch of highway sits the Wellington brewery, one of Ontario's larger and more established craft breweries. Passersby identify the brewery by the cheerful rubber boot sign and tastefully appointed yet rustic taproom.

In the spring of 2020, Cher was looking for a project that her organization could throw itself into, and, backed by the city, she began reaching out to the community. One day she was chatting with the head of Wellington Brewery, and she asked about the sort of waste products the brewery created. What she heard was that the Wellington Brewery — like breweries everywhere — produces a lot of spent grains, leftover from the brewing process. Typically, these spent grains end up fed to animals or composted, but often may just be dumped into municipal landfills where they decompose and contribute to global warming and water pollution. But really, Cher mused, what are spent grains other than a rich supply of organic material and fiber? Maybe someone else in the region could use these?

Soon thereafter, Cher and her team connected the dots between the brewery and one of the region's quirkier start-ups, Oreka Solutions, which produces black soldier fly larvae as a commercial livestock supplement. Located in a disused dairy barn about a twenty-minute drive from the brewery, Oreka had been operating for several years. They collect organic matter from grocery stores that would otherwise go to waste and feed this to the black soldier flies. Of course, it's way more sophisticated than it sounds, and black soldier flies, if they are to grow efficiently, need a carefully constructed and nutritionally balanced diet. But as Cher discovered, the spent grains from the brewery could be easily incorporated into the black soldier flies' balanced diet. And so, Cher and her team, with a few phone calls, connected a few more dots, and within days the brewery was selling their spent grains to the insect farm.

But feeding brewery waste to insects is still not fully circular, and so Cher and team again put on their thinking caps.

Brainstorming with the local networks of entrepreneurs and business owners on what could be done in the region with the insect larvae prompted another series of phone calls and another deal brokered by the early summer of 2020. Cher and her colleagues had arranged for Oreka's insect larvae to be sold as fish food to a local aquaculture operation called Izumi, operating a ten-minute drive from Oreka's base.

Izumi, the fish farm, raises high-quality trout using what they call floating raceway technology. The basic design is simple. Made of fiberglass, the floating raceways are light, narrow tanks that are located within confined ponds and separated from each other by neat wooden walkways. Think of a series of narrow cottage docks, each surrounding the floating raceway container. Because each of the floating raceway tanks is isolated from the pond and from each other it is relatively easy for the fish farmers to control water quality, dissolved oxygen quantity, and temperature while being able to remove what the industry refers to as settable solids — uneaten feed and waste. What's more, since the raceways sit in a larger pond, the overall energy costs for pumping water are very low and comparable with more traditional net pens that might be found in marine fish farms. And, since these floating raceways are placed in confined ponds that are themselves unconnected to local waterways, there is virtually no chance of the farmed fish escaping and out-competing wild fish populations. This means it's also easy to control nutrients, other pests, or diseases from escaping into the environment. All told, floating raceways offer an economically efficient way of producing high-quality fish in an inland environment, thus circumventing many of the traditional criticisms leveled against fish farming, namely that keeping pens of specially bred fish in coastal estuaries or fjords inevitably results in major impact on wild populations as farmed fish, their waste, and the diseases they catch, inevitably work their way into the environment.

Izumi produces two major products. The first is a high-quality trout, normally shipped to high-end restaurants in other cities. But Cher, with a few more phone calls, arranged for some to be shipped more locally, back to Guelph, where a small cluster of restaurants called the Neighbourhood Group operate.

The second product that comes out of Izumi's raceways are the solids scooped up from the bottom. Mostly this is uneaten fish meal and fish manure; in most fish farms this is a waste product that becomes a source of pollution. Much as Cher needed to find a use for the brewery's spent grains, she now made a few more phone calls and found a local potato farmer who was interested in putting these nutrient-dense waste products on their fields. From there it was easy to see what the next step had to be: a few more phone calls, and Cher plus team had arranged for the Neighbourhood Group restaurants to buy the potatoes that were now being fertilized with Izumi's fish waste.

Lenore put her hand up. "OK, let me make sure I've got this straight. By this point, you have the Neighbourhood Group restaurants selling locally sourced beer and using locally produced fish and potatoes. Not only are all these products coming from economically viable companies situated less than an hour's drive away, each of these products themselves is mostly produced by repurchasing the waste products of the other steps in the chain. So, this is basically a closed-loop system operating on a regional scale?"

Cher continued, "Yes, exactly. But we didn't stop there. We asked our team, 'What else does a plate of fish and chips need?' and, 'Can this be sourced locally?'"

Of course, Cher had an answer for this as well: they needed a slice of sourdough bread and some grains for the batter for the fish.

To find these products, Cher and her team went back to the Wellington brewery for a few more of the spent grains not needed by the insects and made two final calls. The first was to a local bakery called Grain Revolution, and the second was to a group of microbiologists from the University of Guelph who had set up a spinoff company called Escarpment Laboratories, which sells yeasts to the local brewing industry and wanted to do something with their spent yeast. Together, the biochemists and the bakers were able to develop a spent-grains sourdough that can go head-to-head with some of the finest artisanal rustic loaves around.

The result was the creation of three circular meals, each of which was sold over the fall of 2020 at three of the Neighbourhood Group's restaurants. First was traditional fish and chips served at the local pub that included a sourdough breading. Second was a smoked trout sandwich on sourdough with hand-cut potato chips featured at the local deli and lunch counter. Finally, was the trout gravlax accompanied by potatoes and horseradish cream and served with a sourdough crostini that was sold at the group's finer dining restaurant. All the meals were a screaming success with consumers and will be fixtures on the restaurants' menus.

Cher concluded her tale of creating a circular food economy on a note of considerable pride. Armed with little more than a cell phone and the need to do something during the pandemic, she made the connections within her community to serve up three gourmet menu options, each of which was locally sourced and used products that would otherwise have ended up as waste.

Later, Lenore and Evan regrouped. "Evan, I like the idea of closed loops on Mars, but maybe they are too complex to plan in advance? Maybe Martians begin with partially closed systems and then fit together the pieces over time? I worry about system collapse if there are no firewalls between processes."

Evan agreed. "I think there are some lessons we can apply from this.

"The first is the easiest to explain but the hardest to apply. The linear take-make-throwaway economy bequeathed to us from the Industrial Revolution causes irreparable environmental problems and needs to be replaced with systems that embrace circularity. This is an immediate imperative when we contemplate what a future Martian community will need as any extraterrestrial food system cannot afford to leak nutrients, water, or carbon dioxide and must be built on the principles of a circular economy."

"Right, and closing the loop on food is also necessary to ensure long-term environmental sustainability here on Earth."

"Agreed. However, the tension this chapter reveals is the lack of financial viability, by which we mean an inability to get the scale, the cost, and the complexity of running closed-loop food systems under control. Furthermore, the examples of aquaponics systems we currently see on Earth do not seem to allow the kind of specialization that brings efficiencies into the food system. So, the second lesson of this chapter is that maybe the appropriate scale at which to consider circular food systems needs to be more regional, where different companies acquire a critical mass in producing their products within a region but then trade with each other. Maybe using a regional approach to this problem, it is possible to have our circular cake and eat it too? This means that for the Martians it may be unrealistic to expect all aspects of the closed-loop system to operate in one container, but perhaps by separating out different functions and services into different domed craters, it will be possible to engineer something that recycles everything?"

"Exactly! And then we get our firewalls back."

What Cher's story demonstrates is that it is possible to devise an almost perfectly circular system that closes the loop on food production, food processing, food consumption, and food waste and embodies all the principles that we are going to need on Earth as well as on Mars. No matter where we end up, the nutrients that go into producing food are far too precious to let them turn into pollution. Instead, we must use examples like Cher's RePURPOSE project as a template to undo some of the problems associated with industrialization and its linear approach to food. If we do this, then we will be able to engage with Wendell Berry's elegant cycles of nature.

CHAPTER 10:
Ballrooms of Mars

PAINTING BASETOWN RED

The pandemic wound on through 2021, and with many fits and starts, life slowly began to change. Canada's vaccination rate meant we were all allowing ourselves a few more freedoms (though Evan kept checking the public health bulletins at 10:30 every morning in a pointless effort to "stay on top of the data"). Lenore began to go to restaurants again and to the beach for picnics. She and Katya even made a trip back to Montreal to see some friends, observing the various health protocols as they went. But many elements were still missing, including one of Lenore's favorite activities: ballroom dancing.

And so, one sultry summer evening, as she emptied beach sand from her shoes, Lenore texted Evan:

"What would a night out in BaseTown look like? Recreation is a critical element of any community, and Mars won't be any different. A girl can't live on Netflix alone."

"Indeed," came Evan's reply, "There has to be more on heaven and Mars than allowed for through terraforming alone!" He then sent a cute shrug emoticon he'd been saving up to impress Lenore with: ¯_(ツ)_/¯

If Lenore moved to Mars, she imagines she would be put in charge of keeping the autonomous controlled fruit production system running. Her day would involve monitoring tunnels filled with strawberries and blueberries basking under LED lights. Her team would be working to perfect smaller avocado plants, and she imagines that after a few Martian months, they would achieve a breakthrough allowing for additional stacked racks of the fruit. Lenore would also be responsible for monitoring citrus production, and (in her daydream) she even made a mental note to report to the chief of agriculture (a very important role in BaseTown) on the new pollinator health program.

Evan, now that he let his imagination run wild, would probably be working on the composting and recycling infrastructure. His dream team in BaseTown would be responsible for keeping the organics system running. This would involve ensuring the collecting and sorting facility was optimized and that the equilibrium in the composting tanks was kept at a constant temperature, venting surplus heat to run electric generators, and ensuring nutrients remained optimal. He'd be calling up to Lenore to get deliveries of unused carbon-rich organics that he'd combined with nitrogen-rich sewage (collected from the human habitat modules) to ensure that the process resulted in fibrous compost and nutrient-dense liquids ideally calibrated for use back in Lenore's growing tunnels.

But when the shifts were over for the day, work would stop. Unlike life on Earth, where everyone seems to be toiling 24-7, folks in BaseTown would be strongly discouraged from working outside designated hours. Life on Mars will always be tough, and burnout would be expensive and dangerous for the whole community. Allowing people to become exhausted would be a luxury only Earthlings could afford.

So, the two of us began imagining what a night out on BaseTown might look like.

In a marked departure from our real Earthling selves, the futuristic Martian versions of Lenore and Evan would turn off their tablets at the end of the shift, brief our replacements, and then rush home to get ready for a night out. We imagine that the two of us, plus our partners Kat and Christine, might be getting together for dinner, just as old friends do. Evan also dreams he would be heading out to a faux Irish pub as he would have reengaged with one of his oldest passions, playing Irish music on his wooden flute. Lenore and Kat would be going dancing with Christine, leaving Evan to his jigs and reels.

But first, Lenore would clean up and slide into a silk dress made from the finest cultured silk. Most fabric on Mars came from the same facilities that made other proteins, and so she could wear silk and wool without silkworms and sheep. Once Katya was ready, Lenore would toss on her leather jacket, a splurge purchase made from the finest cultured leather. After all, the largest dome could get a little chilly after Martian sunset. Katya's handbag would be made from agave leather,[1] created from waste from Mars's only tequila maker. Agave cactuses would likely do surprisingly well in the dry air of BaseTown, saving the farmer a stiff water bill.

By now, we are imagining that BaseTown will have become well enough established that the community has invested in some

---

1   Lenore owns and adores an agave bag.

impressive architecture. The main dome might be a new space that displays the town's success. Funded by the sale of new technology to Earth, the dome might enclose a small crater in three layers of woven glass fabric. Nanites scurry over this hundred-acre space, vigilant for any micro-tears in the covering. Under this dome would nestle the first lake on Mars in a geological age, and lush trees and plants (most fruit-bearing, an effort that stretched Lenore's team to the limit) surround an inner lakeside ring of beaches, restaurants, and public spaces.

In BaseTown, most meals would be consumed in pleasant but standard cafeteria surroundings, where everyone could get food without money exchanging hands. Cooking at home would unlikely be possible beyond a few basics and so the cafeterias are hubs of daily life. But for a special occasion, restaurants would serve luxuries and the main dome would be hopping twenty-four hours and thirty-seven minutes a day.

The present-day Evan and Lenore imagine a future where the four of us would be relaxing over a fine Martian dinner of fresh green salad (Lenore adds a few cubes of cultured tuna ceviche), a red bean rajma curry with a dollop of fermentation-produced sour cream, and finish with mango sticky rice in a coconut syrup. The mango trees would be growing around the perimeter of the main dome, tropical birds flying (somewhat erratically) from branch to branch. Whenever a leaf drops, Evan's internal radar would go off, and he would try not to stare and grow agitated until a small robot (the electronic descendant of the 2020s Roomba vacuum) scoops it up and deposits the precious organic matter back into the recycling system. Mangos are a luxury product, as are the coconuts, but we imagine we might want to splurge a little — in this daydream it is April 12, 2063, the day that falls between Evan and Lenore's birthdays. The red beans were also grown under LED lights, their production even more automated than Lenore's fruit. In some cases, the chef was the first actual person to touch the product.

After dinner, Evan would go to play some Celtic tunes as the rest of the party moves on to sip a drink at a nearby bar. Katya and Christine enjoy a fine Martian red wine from a nearby vineyard tunnel. Lenore sips a bourbon. Though she swears by the traditional corn-distilled liquor, this bourbon is also created through advanced fermentation. Meanwhile, Evan is missing beer (producing barley just for brewing still would likely take too much land). But that isn't too much of a hardship. Every year, Mars improves their fermenting and distilling abilities, and recently one of their Scotches won third place overall in the annual Interplanetary Whiskey Competition. (future-Evan frowned and wondered why the Lunar community always seemed to come out on top. Better water?) Before heading off to our after-dinner plans, the four of us might grab a quick coffee at one of the kiosks. The coffee would also be created using advanced fermentation, and even Lenore had to admit the result is as good as any coffee on Earth; this allows growing space to be shifted into more productive crops. Lenore did insist one of the lakeside groves be given over to cacao trees though, in part because they were so beautiful and aesthetic concerns would have to be considered very seriously by the central planning committee. Keeping the community happy in what is still, after all, a very challenging environment, is central to BaseTown's success. In fact, citizens are so happy and prosperous that we imagine officials from Earth would be running numerous studies to see what they can learn and apply on the home planet, as inequity, alienation, and depression would likely remain serious problems on humanity's birthplace.

Lenore and Katya would change out of their heels and into their dancing shoes and glide onto the floor of a Martian ballroom. There, a late-night band is playing a mix of foxtrot, waltz, and swing. Dancing is somewhat different on Mars, due to the lower gravity, but people there would have learned to amend their style to account for the change in weight and momentum. After a few spirited attempts at Viennese, we imagine Lenore and Kat wandering over to another of the music venues to rejoin Christine and listen to Evan's new folk band. The instruments would be made from precious hardwood grown in the larger corridors of BaseTown, a resource made available only to projects that added to the social good of the community. Evan is jamming so hard that the group rings in the dawn before heading home to enjoy a long snoozy Saturday. The combination of healthy diet, a lack of motor vehicles, work-life balance, and regular exercise and social activity means that we all often forgot we are nearly an Earth-century old. But after all, we wouldn't quite be fifty in Mars years!

WHAT SHALL WE EAT ON MARS?

Our imagined night out on the town represents what might eventually be a special occasion on a new world, but in the early years of a Martian community, what will the day-to-day diet of a Martian look like? Most meals will likely be served communally. Cooking at home is massively inefficient, and in the early days of large cities, including classical Rome and the great cities of China, almost all meals were eaten in cafeterias and inns or on the street. So, let's imagine an average day, when our Martian citizens wander down to the community cafeteria before starting off to work.

For breakfast, the average Martian would probably consume some kind of nutrient-dense bar that would be algae-based. This might be flavored with insect protein but most of the insects would be used to feed fish. Nevertheless, the overall effect is likely to be quite a bit like an Earth-bound granola bar.

Since carbohydrate-rich grain crops take so much space, breakfast bagels are unlikely to feature prominently, but omelets from cellular agricultural bioreactors could be served with a side of hashbrowns made with real potatoes fertilized by composted waste organics collected from the refractories and eating halls. For cooking oil to cook the spuds, algae oil is likely to be in widespread use as it has a neutral flavor and a high smoke point, which makes it a good option for sautéing.

Coffee or tea will be much sought-after luxuries imported from Earth at first, but it's possible to imagine locally produced hot beverages along the lines of instant coffee, and since we already know how to produce synthetic caffeine it isn't hard to imagine that one of the first labs set up on Mars would be a caffeine synthesizer. Advanced fermentation companies have made huge strides even as we wrote this book to synthesize coffee and chocolate, among dozens of other complex products.

The milk or creamer for the hot beverage would almost certainly be a mixture of plant and yeast-fermented dairy fats and proteins. The sweetener would probably be partly fructose (obtained from plants) and partly another yeast-derived protein that we perceive as sweet. Some sweeteners might come from the agave used to make tequila and Katya's handbag. All plants would likely serve multiple purposes.

When it comes to lunch, hyper-fresh salads are almost a certainty given that all BaseTown will be one giant market garden with edible plants integrated into all the living spaces so that every electron and photon in the place is used not only to brighten things up for the humans but also to be collected by plants to fuel photosynthesis.

Along with the salad, we imagine folks eating a 3D-printed fish or chicken cutlet where the proteins are either grown in a bioreactor or are derived by yeasts. These yeasts in turn would feed on the cyanobacteria that are the foundation producers turning the Martian regolith into nitrogen-rich organic matter.

For additional protein, there would likely be some yeast-derived cheese added to the salad. The initial iterations of Martian cheese would presumably be quite homogenous (as they would all use the same proteins and bacterial cultures) but over time, Martian cell-dairy farmers would experiment with creating new microorganisms. Out of these experiments different locally adapted varieties of yeast would give rise to different varieties of cheese that might even create a sense of Martian terroir.

Seasonings and flavor should be relatively easy to come by. NASA's rover *Curiosity* has shown us that there are mountains of salt on Mars, and although this salt would likely need considerable processing before it could be shaken onto Martian home fries, it's even possible to imagine that using Martian table salt might become a status symbol on affluent dining room tables on Earth. Other flavorings can be derived from the plants grown for herbs and spices or cooked up in a lab. The plants growing in the public areas of BaseTown would include herbs and spices and would have the added benefit of sweetening the air in the domes and halls of the community.

Bread products, however, are probably going to be scarce given that it would be difficult and expensive to set up domed habitats capable of growing any quantity of wheat, corn, rice, or barley. Pasta and baked goods will be a luxury to be savored. But at lunch or dinner, there might be potato pancakes or some small pastry made from potato flour.

This relative lack of carbs is okay. On Earth, a major cause of chronic disease is our overconsumption of refined carbohydrates, driving obesity, type II diabetes, and other autoimmune diseases linked with a dysfunctional digestive microbiome. On Mars, we will not be able to afford to produce products that, on Earth, are quite literally killing us. Other more damaging products such as tobacco and hard drugs would likely not be allowed at all, though given humanity's eternal drive to ferment, well, everything, we both imagine alcohol will still play a part in human life on the

Red Planet. A few psychoactive plants and fungi might make the trip too.

At dinner, BaseTownies would sit down to a printed chicken breast and another salad, perhaps washed down with a glass of faux dairy milk or some juiced berries from the vertical farming operations. For a special occasion, the inhabitants might enjoy an actual real-life fish. Fertilized fish eggs could be brought in from Earth in a container that keeps them shielded from harmful radiation and, once on Mars, it should be possible to raise fish as part of the water and waste recycling system, as we saw earlier.

For dessert, sweetener proteins synthesized in the biofoundries might be mixed with synthetic egg proteins and a little bit of (very valuable) flour to create small biscuits that would accompany ice cream or a milkshake (again made with yeast-derived dairy proteins).

For an evening greasy snack, seasoned fried protein balls (salmon, beef, and chicken flavored) and fries could be common. And distilled spirits would likely flow in nightclubs. Martians' experience using yeast to produce dairy products would make them natural brewers, and there would almost certainly be a proliferation of high-test vodka drinks as the most common tipple. Beer, by contrast, would be harder given that growing barley simply to extract sugars would be horrendously expensive. However, if barley could be engineered to produce medical proteins, and if the proteins and the sugars could be separated, then some limited beer might be brewed.

But as greenhouse capacity and vertical farming operations expand, we can imagine that eventually someone would try their hand at growing grapes, and we are confident that the entire community might jostle to buy lottery tickets to be one of the lucky few to sample the first vintage of Martian wine. This wine would probably be enjoyed as a novelty; it is hard to imagine that wine grown without local microorganisms would taste all that great. But like the story of cheese developing Martian terroir, it's possible that advances in our understanding of plants' microbiomes may lead to decent wine. Who knows, maybe a case or two might end up being sent back to Earth on a reverse supply run?

Overall, the Martian diet we foresee is likely to be sensible, tasty, and well-balanced. The biggest difference between what we are imagining and what we eat on Earth today is the lack of livestock products and the relative dearth of simple carbs. But, over time, we think the inhabitants of Mars will not miss these products all that much. Folks there will be, by necessity, eating a diet much more aligned with what nutritionists and national food guides recommend we eat. If we imagine looking around at the people living on BaseTown about a generation after it has been established (and if we assume no great Jamestown-style calamity), we figure that practically no one in BaseTown would be overweight and that commonplace Earthling ailments, such as gluten sensitivity and diabetes, would be nearly unheard of.

CONCLUSION:
Upgrading the Operating System

REWILDING EARTH

Towards the end of writing this book, Evan had a serious moment of nervousness about this whole project. He called Lenore with a sense of existential dread building in the pit of his stomach. "Why should people care about food on Mars? I mean, this whole book is far-fetched, and you and I have spent our entire careers pushing for more efficient, sustainable, healthy food systems here on Earth.

"Lenore, are we wasting time by pretending to do menu-planning for the Space Age?"

"I categorically disagree," Lenore immediately replied. "First, we are going to Mars because that is what we humans do. We push out, we make tools, we make mistakes, and we try again. One of the first people we quoted in this book was Carl Sagan talking about the call of the open road. Another Carl Sagan story involves NASA's *Voyager I* spacecraft. As it was leaving the solar system back in 1990, he convinced mission control to turn the camera back to take one last photograph of Earth. The picture, which is totally worth Googling if you don't know it, shows the Earth as an infinitesimally small fleck. It's called the *Pale Blue Dot* photo.

"Anyhow, upon seeing the tiny speck that was Earth, Sagan commented, 'That's home. That's us . . . everyone you love, everyone you know, everyone you ever heard of, every human being who ever was, lived out their lives.' That's totally true, of course, but another member of the NASA team, Dr. Candice Hassen, said something much more profound. She said, 'I was struck by how special Earth was, as I saw it shining in a ray of sunlight . . . It also made me think about *how vulnerable our tiny planet is* . . .' (emphasis added).[1]

"The real point of our book isn't about feeding Mars as much as it's about protecting Earth. As we've said again and again as we've been working on this project, our industrial food system needs an upgrade to its operating system. And getting to Mars is one way we'll be able to download this upgrade."

So, let's lay it on the line. The two of us hope that we have shown a vibrant Martian community might be possible with technologies that are only a little bit in the future. Nevertheless, our key message is that it is on Earth where this food revolution will have the biggest impact. Many of the tools and technologies described in this book, and designed to sustain the hypothetical Martian community, should immediately find their way into our economy and become incorporated into farming and food systems here on Earth. And as these new tools roll out, they should have the remarkable effect of both reducing the environmental footprint of terrestrial

---

[1]   Here is the NASA article that quotes Sagan and Hassan (though the Sagan quote is pulled from the introduction to his book *Pale Blue Dot*): https://www .nasa.gov/mission_pages/voyager/voyager-20100212.html

food systems as well as providing billions of Earth dwellers a much healthier diet than we now find.

Our hope is that through books like this, we and others will throw down a trail of breadcrumbs that other scientists, politicians, and industry leaders will follow until we reach a point where it will seem that the old twentieth-century ways of doing agriculture look prehistoric. We hope that in the not-too-distant future, we all look back at the early 2020s as what we'd like to call Peak Farmland, Peak Meat, and Peak Water.

Our food system today is like a tidal wave flowing over the world. Ten species[2] dominate about 39 percent of the planet — 14 percent dedicated to cropland and another 25 percent held for forage and grazing (that is almost entirely devoted to cattle). These numbers *cannot* keep growing, and with the world poised to need to feed approximately ten billion humans, it is impossible that we continue to press against natural ecosystems. We cannot meaningfully expand our cropland any further, and we must shrink our pastureland. The number of cows, pigs, and chickens on the planet needs to drop dramatically.

Agricultural technology advances should allow the floodwaters to begin falling back. Better quality seeds are already spreading around the world helping to protect harvests against droughts. Smart tractors are putting fertilizer down more efficiently and our enhanced understanding of both nanotechnology and the microbiome are creating new tools to produce food with reduced inputs. A raft of new ways of producing proteins and fruits and vegetables give us hope that our agricultural system can be rejigged to significantly reduce its impact on the planet. Doing this should also help keep healthy food affordable.

2　Wheat, corn, rice, soy bean, palm, sugar beet, potato, pigs, cows, and chicken.

But it's not happening fast enough. The world is already on its way to fail the not-very-ambitious climate change targets set by governments over the past ten years, and already the resulting problems are mounting everywhere. Nonetheless, we've all just been through the most amazing two years where COVID-19 has shown us both the brilliance and the folly of our species. In late 2019, COVID was essentially unknown. Yet by mid-2021, most of Canada is fully vaccinated and there are not only numerous vaccines but also a growing range of antivirals to choose from. Before 2020, most medical researchers would have said it takes ten years to bring a new vaccine to market yet here we are: most of the Developed World is vaccinated and while there is still much work to be done, and we hope that future variants won't emerge and keep setting us back, we are in a much better place to fight infectious diseases in general than ever before. To quote Matt Damon's character in the movie *The Martian*, "we scienced the shit out of [this problem]."

So maybe when it comes to transforming our food systems, we need to aim for a moonshot that is sexier and grabs the imagination better than simply saying "we need more sustainable food systems." And maybe this is what a Martian mission could do — give us a big audacious goal and set us on a path to reach that goal.

It is ironic that the two of us sustainable foodies are imagining flying to a different planet — a planet that as far as we know is completely bereft of life — to save the rich natural heritage here on Earth. But we feel that the 2020s represent a critical juncture in human history. Today, our relationship with planet Earth is akin to a burning library in ancient times or like the lost Mayan texts that were put to the flame by conquistadors. Today, the extinction of major parts of planet Earth's biological diversity, unique in the universe as far as we know, is just like the great library in Alexandria that vanished in a puff of ancient smoke. We are letting our one and only planet burn down around us. The only hope for humanity is to use every tool in our formidable technological arsenal and do everything we can to save our planet from us.

A second irony of this book that the two of us have come to realize (only lately) is that it is the genetic heritage of the Earth that has allowed us to become so lazy with how we feed ourselves. It is the abundance of land, water, soil, and species — an evolutionary heritage that Mars will never have — that has allowed us the luxury of developing food and farming systems that are staggeringly inefficient.

A third irony: none of this is the fault of farmers, yet it is likely farmers who will have to adapt the most. All over this planet, farmers toil with extraordinary energy, discipline, and ingenuity under a political and economic system that refuses to see the value of their work. But the old ways of doing agriculture have reached their limits and we need a catalytic moment to galvanize us to reimagine a new way of doing things.

On Earth, we have been able to push off many of the negative costs of producing food onto the backs of poorly paid labor, the suffering animals, the environment, and onto future generations. The ability of our planet to absorb abuse has allowed us to create the profligate industrial food system. On Mars, there can be no such lack of discipline. We must be incredibly efficient right from the get-go.

DON'T BET AGAINST THE FOOD SYSTEM

We are in a race to save planet Earth, but neither of us wants to conclude on an apocalyptic tone, and both of us are optimists. And we know that for centuries, "learned" people like us have been betting against the food system. In 1798, Thomas Robert Malthus argued that demand for food caused by population growth would necessarily outstrip supply and drive humanity to inevitable famine. In this,

Malthus bet against agriculture in such a spectacular fashion that an entire brand of ecological pessimism — called Malthusianism — bears his name.

In 1980, something similar happened when an economist made a literal thousand-dollar bet against an ecologist. Julian Simon, a business professor at the University of Maryland, wagered Paul Ehrlich, an ecologist at Stanford University, that the cost of raw materials would fall over the decade. Ehrlich chose a set of raw materials and the two agreed to reconvene on Sept. 29, 1990. If prices rose (a sign of scarcity), Ehrlich won. But if they fell (a sign of abundance), Simon would come out on top. The reason for the bet related to each man's worldview. Simon was a strong proponent that innovation and technology allow us to overcome limits to growth. Ehrlich observed the world's environmental problems and argued population growth would result in famine, scarcity, and ruin.

Today, with climate change looming, heat domes shattering temperature records in Canada, and California's agricultural land struggling to remain productive, many food system experts are laying down the same bet. As the ice melts, fires burn, and storms rage, it's hard to imagine how the current agricultural system will survive, but will its decline take humanity with it? Or will we pivot into something different? Hence, the key take-home message of this book is a prediction that through a catalytic project like a Mars mission, we may end up in a future more like the one envisaged by the optimistic Julian Simon.

But there are important nuances. Technology alone will not be enough to strengthen the world's food systems. Most of the world's 570 million farms are small-scale and family-run and won't have access to most innovations. It is possible that certain innovations could even drive the decline of small- and midsize producers, leading to upticks in unemployment or poverty. Similarly, if the economic system doesn't pay workers a fair wage, ignores animal welfare, and externalizes environmental costs, then the technologies described in this book won't help. Ensuring we end up on the right side of this equation is where policy comes in. Government regulations must put a price on things like greenhouse gas emissions and water pollution so that farmers who are good stewards of the environment are rewarded.[3]

When Simon and Ehrlich reconvened in 1990, it turned out that the economist had triumphed. All the resources Ehrlich identified declined in price over the 1980s. Simon crowed about the role of ingenuity and innovation. Ehrlich grumbled he'd chosen badly and that a recession in 1990 artificially dampened prices. Both academics were partly right and partly wrong. Ehrlich underestimated the innovation Simon celebrated. But Simon did not appreciate the importance of strong policy to protect labor and environment.

As we look at the twenty-first century, a century that threatens massive disruptions but also promises huge innovations, we need two things. We must capitalize on the technology that can help us change the way we produce food. And we can never forget the importance of public policy to ensure there's a fair price put on things such as biodiversity, climate change, human labor, and animal welfare. If we embrace both principles, there is a very real chance that we will be able to bring the price of producing healthy food down without destroying the ecosystems we all depend on for life.

The thought experiment of developing self-sustaining human habitation on Mars is our attempt to channel a bit of the post-vaccination scientific optimism we are both feeling. And here we have presented a vision of how we can play to one of humanity's greatest strengths — the development and deployment of technology — to one of the great global challenges facing the twenty-first century.

---

3   Lenore and Evan have both written extensively about this in other places, so if you are interested in this side of their thinking, drop either of them a line.

A RETURN TO THE BOAT PLACE

In 1845, the Franklin expedition left the UK searching for the Northwest Passage and sailed into oblivion. Before they departed, the ships were loaded with 36,000 pounds of biscuit, 136,000 pounds of flour, over 1,000 pounds of pemmican, 60,000 tons of salted meat, and — critically — 33,000 pounds of tinned meat. It was enough for three years.

As the months of no news dragged on, Sir John's wife — Lady Franklin — began to raise the alarm. But her concerns were dismissed by the officers from the Admiralty Office in Whitehall. The food carried by Franklin's ships would still be nourishing the crew. There was still time. Plus, Franklin and his men were the finest of the British Navy, and they were backed up with cutting-edge technology. Their boats were recently outfitted with better engines. They carried tinned food — a modern miracle that kept men healthy and strong during the most trying of conditions.

"We mustn't panic," the officers must have patronized the distraught (and indomitable) Lady Franklin.

But as the calendar turned over into the New Year of 1848, and with still no word from the expedition, even the brass from the British Admiralty must have begun fearing the worst. Franklin's men were now almost certainly running low on provisions. The tinned goods must be used up. Something had gone awry.

But by 1848, it was too late to do anything but look for the wrecks and try to account for the loss. Later that year, the Admiralty and Sir John's widow mobilized the first of the search expeditions that would brave the Arctic to try and discover her husband's fate. One of these search parties included the Scottish surgeon John Rae, then in his thirties and already a seasoned Arctic explorer having worked as a surgeon for the Hudson's Bay Company.

Rae's accounts of his time exploring the Arctic are dry and clinical, yet reading them today evokes months of grueling effort, inhumanly harsh conditions, and barren landscapes. But Rae, unlike most members of the Franklin expedition he was searching for, knew what he was up against. Among his many accomplishments, he was the first European to live off the land during the entirety of an Arctic winter, eating only what he could gather and hunt. His career would see him map the final missing link in the Northwest Passage, discovered through a thousand kilometer hike along the coast.

To the modern reader, Rae's record of survival seems superhuman, and among European settlers at the time, he was widely acknowledged as one of the very best in terms of hunting, snowshoeing, and winter survival. But he owed his skills to the local Indigenous communities he lived and worked with. He was unusual compared to most men of his generation in that he considered Indigenous North Americans his equals. His respect for local knowledge is what allowed him to survive and gave him the advantage over other explorers.

The initial searches for Franklin's fate were dead ends. But Rae persevered, and in the spring of 1854, he met a group of Inuit who conveyed that, a few years earlier, some European sailors had starved in the region, cannibalizing each other in their final desperate throes. The Inuit offered proof — some trinkets they had picked up from the tragedy, including a silver plate engraved with Franklin's name. Rae traded for these and told them he'd buy anything else they could find and then, with the weather turning against him, wisely hurried away to file his report and present his proof.

What happened next was both shocking and predictable. The Admiralty, Franklin's widow, and European society weren't willing to accept that one of their own could have failed so spectacularly. The arrogance and racism of the time meant that John Rae's report was dismissed by some and met with scorn and derision by others.

It was unthinkable that Franklin's men would have been reduced to cannibalism. It was unthinkable that things could have gone so wrong. Apparently, Lady Franklin even enlisted the support of Charles Dickens who wrote a pamphlet in 1854 called *The Lost Arctic Voyagers* where he implied the Inuit had probably killed the brave soldiers and sailors. Of course, Dickens had no proof of this, but in a passage that highlights the bigotry of the time, he argued that no one could prove a "sad remnant of Franklin's gallant band were *not* set upon and slain by the Esquimaux themselves," adding that he "believe[d] every savage to be in his heart covetous, treacherous, and cruel . . ."[4]

Of course, the unthinkable had occurred and the Inuit were right. It took Captain Francis McClintock's journey in 1859 and the grisly discovery at the Boat Place, before the hubris of Franklin's expedition began to sink in.

Modern forensics and anthropology have revealed much about what doomed Franklin's expedition. We now know that the ships were ill-equipped to handle the ice and that the expedition was unlucky as the winters of the mid-1840s were unusually hard. We know that the tinned food they took along was soldered with lead and this may have given them lead poisoning. Scurvy would have added to the list of problems they had to contend with, along with hypothermia, frostbite, and snow blindness.

Historians now also suspect that the seemingly miraculous tinned food had been processed in a hurry and by a company with a spotty record of quality control. Much of this food probably spoiled well before its best-before date, and this meant the crew would have been running low on calories much earlier than anticipated. As they starved, the men probably grew desperate and ate the putrefied

---

4   Full text of the pamphlet can be found here: https://victorianweb.org/authors
/dickens/arctic/pva342.html

messes in the badly sealed cans. At best, this would have caused food poisoning — further weakening already damaged immune systems. Death by botulism was also quite likely.

But John Rae was different. He knew, or learned, many of these things, especially the importance of knowing how to use the local resources effectively. And European society lambasted him for it and ridiculed his reports on Franklin's fate.

The lesson of the lost Franklin expedition, the Inuit, the Boat Place, and John Rae is pertinent to our conversation about Mars. As humanity looks to start a new age of discovery, we need to remember the sin of hubris. We should not put all our faith in technology. We must remember humility and collaboration are key ingredients in success, and we must strive to work within the constraints of existing systems rather than imposing an alien system on an indigenous one.

On Mars, this will be vital. Of course, intrepid Martian explorers will need to take food along as a backup and for the long journey. But if a Martian community is to survive, it will also need to work with what Mars has to offer locally, including atmospheric carbon dioxide, nitrogen, and argon, plus the phosphorus, water, and oxygen found in the ground.

If we don't do this, then it is practically certain that there will be an equivalent of the Boat Place on Mars. Some red-hued crater where a rover containing a couple of brave souls got caught in a dust storm and broke down. Unfortunately, by the time the search party was dispatched, the astronauts will have starved, asphyxiated, or died of thirst.

But the real lesson of the Boat Place — and the real message of this book — is that by closing our manuscript with a reflection on the hubris of the last Age of Discovery, we are only partly talking about our current obsession with rocket ships and extraterrestrial communities. What we are really doing is talking, of course, about how we can all live better here on Earth by respecting limits, being careful and efficient with inputs including human labor, trying to keep our systems thrifty and compact, and reusing all waste.

As the two of us have written this book, we feel that we have been toiling under two shadows. The first has been cast by a cloud of rocket fuel, left behind as the current batch of space agencies and billionaires blast off for outer space. The feats of engineering demonstrated by SpaceX's self-landing rockets, and the raft of Martian rovers that have landed on the Red Planet in 2021 are astounding. And our sense of wonder at the miraculous engineering is twinned with a sense of bemusement at the vaingloriousness epitomized by the stunningly phallic rocket the world's second-richest man chose as his first ride off-planet. Our sense is that today we are observing a similar mixture of optimism and arrogance that comes across in the nineteenth-century accounts written by the colonial explorers as they set off from Europe to conquer new lands.

The second shadow is, of course, the pandemic. While there will be an entire generation of historians who spend their careers trying to make sense of COVID-19, for now, let us just reference three conflicting emotions. First, in our hot, crowded, unsustainable world, global catastrophes such as the pandemic are inevitable and will likely come raining down on us during the twenty-first century. Second, our ability to develop innovative scientific solutions, as witnessed by the vaccines, is astounding. Third, our ability to govern ourselves sensibly and proactively is terrible.

The cliché suggests that history keeps repeating itself. This may be true, but today climate change and population growth mean the stakes are, for the first time in history, truly global. So, if our global society wants to collectively avoid the fate of Franklin's men — if we all don't want to die on a barren and desolate rock — and if we want to keep our own planet vibrant enough to see our children and grandchildren have a decent future, we need to do at least two things. We must embrace both the tools and techniques of science and match this with a humbler understanding of what the past can teach us about how to apply these tools. If we do this, perhaps we will see a way of charting a better path into the future as our species contemplates moving to the stars.

Suggested Reading

For the interested reader, the two of us would like to offer a few additional resources that might be useful. First, we can't help but highlight that both of us have written a lot about history, the world food system, and related topics. As such, we'd like to recommend Lenore's *Lost Feast* (ECW Press, 2019) and *Speaking in Cod Tongues* (University of Regina Press, 2017) and Evan's *Empires of Food*, *Beef* (William Morrow, 2008) and *Uncertain Harvest* (University of Regina Press, 2020).

In terms of non fiction about setting up shop on Mars, there are plenty of books on the market. Some of our favorites include Robert Zubrin's *The Case for Mars* (Simon & Schuster, 2012) and Stephen Petranek's *How We'll Live on Mars* (Simon & Schuster, 2015). And of course, there is Carl Sagan's *Pale Blue Dot* (Ballantine Books, 1994). It is in the realm of fiction, however, where things get truly exciting. Kim Stanley Robinson's Mars Trilogy is a classic everyone should read (first published in the 1990s but reprinted many times since). A more recent novel that Evan loves is Emma Newman's *Before Mars* (Ace Books, 2018).

When it come to the more technical aspects of the science, there are a huge number of first-rate popular science titles on pretty much all the topics covered in our book. For example, there are lots of good books about nanotechnology, including Sonia Contera's *Nano Comes to Life* (Princeton University Press, 2019). Of course, the science is moving so fast in these fields that it is the academic work that contains most of the cutting-edge stuff. Similarly, most of the published work on the soil microbiome, along with research attempting to harness algae to create autonomous food systems, is mostly academic and not much of this science has entered the public sphere. With that said, there are lots of good nonfiction books about the human microbiome, such as Guilia Enders's *Gut* (Greystone Books, 2018).

When it comes to greenhouses, vertical farming, and doing agriculture in controlled environment settings, probably the best book we have come across is the Toyoki Kozai, Genhua Niu, and Michiko Takagaki–edited volume *Plant Factory* (Academic Press, 2019). *Plant Factory* provides a comprehensive overview of the technologies around indoor growing and vertical farming (but is probably too technical for most people). More generally, if readers are interested in delving into topics related to futurism, then there is Jonathan Keats's book on Buckminster Fuller, *You Belong to the Universe* (Oxford University Press, 2016).

In terms of better understanding the relationship between society and the environment, such as we explored in the chapter "Grass 2.0," David Montgomery's *Dirt* (University of California Press, 2012) takes the reader on a survey of how we depend on Earth's topsoil.

Popular science authors are now writing about the science of cellular agriculture and new ways of producing protein. Notably, interested readers might find the following two books interesting:

Chase Purdy's *Billion Dollar Burger* (Portfolio, 2020) and Benjamin Wurgaft's *Meat Planet* (University of California Press, 2019).

Finally, on the topic of how our food systems drive climate change and other environmental problems, we both really enjoyed Bill Gates's *How to Avoid a Climate Disaster* (Knopf, 2021) and Jessica Fanzo's *Can Fixing Dinner Fix the Planet?* (Johns Hopkins University Press, 2021).

Acknowledgments

A s always, books are team efforts. In addition to the people interviewed and named in the text, we have been helped by supremely supportive and encouraging colleagues from all over the planet but especially at our home universities of University of the Fraser Valley and Guelph. A lot of the insights the two of us drew on came from various research projects we have run over the past five years that were funded by organizations such as the Social Sciences and Humanities Research Council and the Canada First Research Excellence Fund–supported project Food from Thought. Thanks to our agent Tim Travaglini, ECW staff, and editor at ECW Susan Renouf. Thanks also to Dr. David Fraser from the University of British Columbia for his superbly useful comments and edits on early drafts.

This book is also available as a Global Certified Accessible™ (GCA) ebook. ECW Press's ebooks are screen reader friendly and are built to meet the needs of those who are unable to read standard print due to blindness, low vision, dyslexia, or a physical disability.

At ECW Press, we want you to enjoy our books in whatever format you like. If you've bought a print copy, just send an email to ebook@ecwpress.com and include:

- the book title
- the name of the store where you purchased it
- a screenshot or picture of your order/receipt number and your name
- your preference of file type: PDF (for desktop reading), ePub (for a phone/tablet, Kobo, or Nook), mobi (for Kindle)

A real person will respond to your email with your ebook attached. Please note this offer is only for copies bought for personal use and does not apply to school or library copies.

Thank you for supporting an independently owned Canadian publisher with your purchase!

This book is made of paper from well-managed FSC® - certified forests, recycled materials, and other controlled sources.